Remote sensing and geographic information systems (GIS) are inherently linked technologies. Together, they form a powerful tool to measure, map, monitor, and model resources and environmental data for both scientific and commercial applications.

This book brings together work by leaders in the field to address improved techniques, applications, and research in integrated remote sensing and geographical information systems. Among the topics covered are image rectification, change detection, visualization, computer modeling, local land use planning, requirements of GIS in global change research, and a look at what the future could hold.

Remote sensing and GIS play a key role in studies of water resources, air quality, topography, land cover, and population location and dynamics, and specialists working in these areas at the local, regional, and global levels will find this book a valuable and up-to-date information source.

Integration of Geographic Information Systems and Remote Sensing

TOPICS IN REMOTE SENSING

Series editors
R.E. Arvidson and M.J. Rycroft

Integration of Geographic Information Systems and Remote Sensing

Edited by

Jeffrey L. Star

University of California
Santa Barbara

John E. Estes

University of California
Santa Barbara

Kenneth C. McGwire

Desert Research Institute
Biological Sciences Center

CAMBRIDGE
UNIVERSITY PRESS

CAMBRIDGE UNIVERSITY PRESS
Cambridge, New York, Melbourne, Madrid, Cape Town, Singapore,
São Paulo, Delhi, Dubai, Tokyo, Mexico City

Cambridge University Press
The Edinburgh Building, Cambridge CB2 8RU, UK

Published in the United States of America by Cambridge University Press, New York

www.cambridge.org
Information on this title: www.cambridge.org/9781521158800

First published 1997
First paperback edition 2010

A catalogue record for this publication is available from the British Library

Library of Congress Cataloguing in Publication Data

Integration of geographic information systems and remote sensing/
edited by Jeffrey L. Star. John E. Estes, Kenneth C. McGwire.
p. cm.– (Topics in remote sensing: 5)
Includes bibliographical references.
ISBN 0-521-44032-7 (hbk)
1. Geographic information systems. 2. Remotesensing. I. Star,
Jeffrey. II. Estes, J. E. III. McGwire, Kenneth C. (Kenneth
Christian), 1962- . IV.Series.
G70.212.157 1996   96-14027
910'.28–dc20   CIP

ISBN 978-0-521-44032-5 Hardback
ISBN 978-0-521-15880-0 Paperback

Additional resources for this publication at www.cambridge.org/9780521158800

This book is dedicated to the memory
of Jeffrey Lewis Star, a talented scientist
and a loving husband and father

CONTENTS

CREDITS

Dr. Gassam Asrar
Office of Mission to Planet Earth
NASA Headquarters, Code YS
Washington, D.C.

Dr. Dave Cowen
Department of Geography
University of South Carolina
Columbia, South Carolina

Dr. Manfred Ehlers
Institut für Strukturforschung und Planung in
  agrarischen Intensivgebieten (ISPA)
University of Osnabrück-Vechta
Vechta, Germany

Dr. John E. Estes
Information Science Research Group
Geography Remote Sensing Unit
University of California, Santa Barbara

Dr. Nick Faust
Georgia Technology Research Institute
Atlanta, Georgia

Dr. Timothy W. Foresman
Department of Geography
University of Maryland Baltimore County
Baltimore, Maryland

Dr. Michael Goodchild
National Center for Geographic Information and
  Analysis
University of California
Santa Barbara, California

Joanne Halls
Research Planning, Inc.
1200 Park Street
Columbia, South Carolina

Dr. John R. Jensen
Department of Geography
University of South Carolina
Columbia, South Carolina

Dr. Kenneth C. McGwire
Biological Sciences Center
Desert Research Institute
Reno, Nevada

Dr. Thomas L. Millette
Department of Geography
Mount Holyoke College
South Hadley, Massachusetts

Sunil Narumalani
Department of Geography
University of Nebraska
Lincoln, Nebraska

Amitabh Saran
Department of Computer Science
University of California
Santa Barbara, California

Dr. Terence R. Smith
Department of Geography and Department
  of Computer Science
University of California
Santa Barbara, California

Dr. Jeffrey L. Star
Information Science Research Group
Geography Remote Sensing Unit
University of California, Santa Barbara

Dr. Jianwen Su
Department of Computer Science
University of California
Santa Barbara, California

PREFACE

Remote sensing and geographic information systems (GIS) are inherently linked technologies – technologies that in many ways share common historical roots. Though the time frame of intense digital GIS development lagged behind digital remote sensing by more than a decade, it can now be argued that the applications of GIS have far outstripped the applications of digital remotely sensed data in the marketplace. Perhaps this lag in historical development was primarily the result of the lack of available digital data that could be input into GIS. Perhaps this lag was conditioned by the fact that although the development of remote sensing systems was heavily conditioned by the Federal government, the development of GIS was primarily in the hands of private enterprise. Remote sensing systems essentially evolved out of classified technologies and were originally justified and developed to meet U.S. Government data/information needs. GIS, on the other hand, came out of academia and private entrepreneurs. Development of GIS began slowly, but the pace of their growth began to pick up in the 1980s.

The number and range of applications that use GIS have exploded in the last few years. Remote sensing, however, appears to be in a somewhat more shaky position. While some in the science community continue to criticize the Earth Observing System (EOS) and its data and information system (EOS/DIS), a number of companies have sought licenses to fly 1-meter resolution sensor systems on polar orbiting platforms. In the next few years the first of these systems is scheduled to be placed into orbit. Only time will tell if these first systems are too far ahead of the curve, but if they are it will not be by far. The time for such systems is coming. The need for the data that such systems can produce is overwhelming. The imperative for the information that these systems can generate is tremendous.

The problems of this planet require data/information in a timely fashion at scales from local to global. Increasing population, agriculture, food and energy security, conversion of land surface cover, climate change, public health, and cultural conflicts are examples in which information is required at a global scale. Retail shop location, automobile navigation systems, 911 locators, and precision agriculture are examples at the local end of the scale spectrum in data/information needs. Yet, by-in-large, the data required to address these needs does not exist. Recent conferences and workshops have validated this statement. From state and local representatives in the United States to national and international governmental officials, there is agreement that data on important attributes such as population location and dynamics, topography, land cover, water resources, and air quality are inadequate, or in some areas they simply do not exist. Remote sensing is the only practical means to acquire much of the data needed to address the wide variety of challenges we face across all scales. To be effectively analyzed and employed, remotely sensed data must be combined with other data/information. The most effective and efficient way to accomplish this is within the context of a geographic information system. For GIS to be most effective, they need to contain accurate, up-to-date data. The best way to update a number of key GIS data planes is through the analysis of remotely sensed data. As stated above, the two technologies are inherently linked.

The idea for the book of readings grew out of the National Center for Geographic Information and Analysis (NCGIA), Initiative 12: "The Integration of Remote Sensing and GIS." The authors gratefully acknowledge NCGIA support in this effort. We particularly acknowledge the support from NCGIA's grant from the National Science Foundation, NSF SBR88-10917. We also acknowledge support from other agencies including the U.S. Geological Survey, the Environmental Protection Agency, and the National Aeronautics and Space Administration (NASA). We also acknowledge support from Edgerton, Gerhardt, and Gershon (EG&G). We give a special note of thanks to the support provided for this effort by NASA grant NAGW-1743 to the Information Science Research Group at the University of California, Santa Barbara.

There are also many individuals who should be thanked for their contributions to this effort. Mike Goodchild, Terry Smith, Joe Scepan, and Karen Kline come quickly to mind. Others include Alex Tuyahov, Dixon Butler, Don Lauer, and Al Watkins. Shelby Tilford also deserves a great deal of credit. Ray Brynes, R.J. Thompson, Martha Maiden, Nancy Maynard, and

Jerry Garegagni do too. Many of these people may not realize it, but they have provided inspiration that has kept us going at times. So too have Tim Foresman, Claire Estes, and Toni Star in many very real ways. Finally, this book has taken longer than it should have to produce because of the passing of the dynamo that was Jeffrey Star. We miss his energy, his infectious enthusiasm, and his intellectual insight. This book is dedicated to his memory.

1

Integration of geographic information systems and remote sensing: A background to NCGIA initiative 12

Jeffrey L. Star, John E. Estes, and Kenneth C. McGwire

1.1. Introduction

The National Center for Geographic Information and Analysis (NCGIA) was awarded by the National Science Foundation (NSF) to a consortium of universities in August of 1988. NCGIA was established to remove impediments to the broader application of geographic information systems (GIS) and geographic analysis. The three universities forming NCGIA are the University of California, Santa Barbara (UCSB), the State University of New York (SUNY) Buffalo, and the University of Maine at Orono. NCGIA conducts activities in three areas: research, education, and outreach. Education and outreach activities involve curriculum development, participation in workshops and conferences, and cooperative activities with public and private organizations. Research at the National Center to date has been primarily centered around conducting research initiatives. Twelve research initiatives were originally proposed in the consortia proposal. These initiatives, which would be conducted over a three-year period, are:

I-1. Accuracy of Spatial Databases,
I-2. Languages of Spatial Relations,
I-3. Multiple Representations,
I-4. Use and Value of Geographic Information in Decision Making,
I-5. Architecture of Very Large Spatial Databases,
I-6. Spatial Decision Support Systems,
I-7. Visualization of the Quality of Spatial Information,
I-8. Expert Systems for Cartographic Design,
I-9. Institutions Sharing Spatial Information,

I-10. Temporal Relations in GIS,
I-11. Space-Time Statistical Models in GIS, and
I-12. Integration of Remote Sensing and GIS Technologies.

Through the fall of 1990, Initiatives 1 through 6 were held. Based upon the widespread interest within the Federal establishment and private industry, the NCGIA Board of Directors recommended that I-12 be moved forward in the initiative sequence. A primary objective of the NCGIA initiative process is to define a prioritized research agenda, and I-12 focuses upon identifying those specific areas where research is required to remove impediments to the integ ration of remote sensing and GIS- in essence, impediments to what some have termed an integrated geographic information system (IGIS).

Earth observing sensors on aircraft and spacecraft provide researchers, environmental planners, and resource management and policy making personnel with powerful tools for producing spatial and temporal information. Remote sensing technology has been used for over a century for a variety of environmental applications. From the acquisition of imagery using tethered balloons in the 1850s to the satellite platforms of today, the amount and types of data has increased dramatically. Figure 1.1 depicts the increasing complexity of remote sensor technology. By developing a temporal sequence of remote sensing data products, information for monitoring and management can be provided to the decision- and policy-making processes. Geographic information systems provide users with tools for effective and efficient storage and manipulation of remotely sensed and other spatial data for scientific, management, and policy oriented information. These two technologies together may provide excellent capabilities for measurement, mapping, monitoring, and modeling for a variety of scientific and commercial applications. A synergism exists in which remotely sensed data can provide the necessary information to keep a GIS database up to date, while GIS may improve the extraction of information from remotely sensed data.

This chapter provides a background to NCGIA Initiative 12. The objectives of the initiative and the process involved in the development of this research

Figure 1.1. The developing complexity of remote sensing analysis

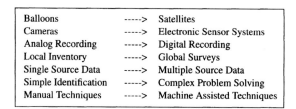

Balloons	----->	Satellites
Cameras	----->	Electronic Sensor Systems
Analog Recording	----->	Digital Recording
Local Inventory	----->	Global Surveys
Single Source Data	----->	Multiple Source Data
Simple Identification	----->	Complex Problem Solving
Manual Techniques	----->	Machine Assisted Techniques

initiative are discussed. Results are then discussed within each of the five major areas of emphasis selected by initiative participants. The chapters of this monograph represent research efforts related to NCGIA I-12. As with all research activities, the material in this monograph represents a slice in time. Hopefully some of the challenges in integrating remote sensing and GIS technologies will be resolved by the time you read this; however, it is likely that other issues will remain relevant for years to come.

1.2. Research issues

The goal of I-12 was to create a prioritized research agenda to remove impediments to the fuller integration of remote sensing and geographic information systems. NCGIA I-12 leaders held a planning meeting in May of 1990, during which they identified general areas of emphasis for the initiative and potential participants for a subsequent specialist meeting. The subject areas were error analysis, data structures and access to data, data processing flow and methodology, man–machine interaction, hardware environments, and institutional issues. Individuals were selected to outline potential research topics within each of these areas, and draft position papers were developed as starting points for discussion at the specialist meeting. A second planning meeting was held at the Stennis Space Center, Mississippi in August of 1990. Attendees continued to flash out issues required for a research agenda, and the list of participants was further refined to ensure a broad cross section of discipline interests as well as fundamental research and application areas. At this time the five areas of emphasis were finalized for the development of position papers. It was decided to combine the man–machine interaction and hardware environments topics from the previous meeting into a new topic called future computing environments. The resulting five areas of emphasis were:
(1) Error Analysis,
(2) Data Structures and Access to Data,
(3) Data Processing Flows and Methodology,
(4) Future Computing Environments, and
(5) Institutional Issues.

The NCGIA Initiative 12 specialist meeting was then held at the EROS Data Center in Sioux Falls, South Dakota in December of 1990. The charge presented to meeting participants was to review and critically discuss the draft discussion papers that had been developed in each of the five areas. Presentations were made of the draft material for each of the five topics. Each presentation was followed by remarks from discussants. The two discussants assigned to each paper were drawn from government, academic, and private sector backgrounds. The initiative organizers felt that the discussant format would stimulate participation among all meeting attendees. These presentations were followed by open discussion among participants. Attendees were asked to validate or reject specific research issues and to give a preliminary prioritization of issues raised in the presentations and discussions. On the third and final day, meeting participants broke into working groups aligned with the five topic areas. An executive session then examined the steps required to prepare for technical sessions on I-12 at the ASPRS/ACSM meeting in Baltimore in March 1991 and to plan revisions of the five presentations as a submission to *Photogrammetric Engineering and Remote Sensing*. Additional discussion revolved around what was to eventually become this monograph.

The following material in Sections 1.2 through 1.6 summarizes findings from each of the five working groups that were published in the June 1991 issue of *Photogrammetric Engineering and Remote Sensing*.

1.3. Error analysis

Issues raised by the Error Analysis working group were documented by Lunetta et al. (1991) and are summarized here. NCGIA Initiative 1 already dealt with issues related to accuracy in GIS analysis so attention in this initiative was given to understanding the errors that can impede the use of remote sensing within a GIS context. Remote sensing provides a significant source of input to GIS analyses. As such, GIS users are concerned about the types and magnitudes of errors in data products derived from remotely sensed data. Participants felt that considerable research is still required before uncertainty associated with the integration of remote sensing and GIS can be adequately quantified and translated into subsequent decision-making processes. Research requirements were identified in the following subtopics:
(1) Data Acquisition,
(2) Data Processing,
(3) Data Analysis,
(4) Data Conversion,
(5) Error Assessment, and
(6) Final Product Presentation.

Based on a review of these subtopics, the Error Analysis working group proposed several priority items. It was suggested that research was required to assess and propose error reporting standards and documentation of data product lineage. Further, it was recommended that by developing a classification of types of error and standardized methods of error reporting that data transfer standards such as the spatial data transfer standard (SDTS) might be extended to ensure that this information is always passed on to subsequent data users.

Current remote sensing error assessment procedures are adapted from statistical procedures that were not developed for spatial data. Although a wide range of peer-reviewed papers on accuracy assessment have been published (see Fenstermaker, 1994), the working group identified a need for improved techniques for assessing the spatial structure of error in remote sensing data products and consideration of this uncertainty in subsequent analyses. Such efforts would include maintaining a connection between mapped units and the by-class accuracies derived from error assessment, quantifying the effect of mixed pixels along polygon boundaries, and promoting the use of continuous confidence surfaces created by statistical classification techniques. The need for "ground truth" to assess the accuracy of remote sensing data products is well established . However, it is often infeasible to conduct exhaustive ground truthing in the field, so a multistage approach using aerial photography is typically used. The working group identified a need for research on the levels and characteristics of accuracy that may be derived from various accuracy assessment methods.

Whereas image classification is usually performed by per pixel labeling of raster data, in many instances the desired result of image classification is a vector-based map of polygons. In this case one may want to simplify or generalize the per pixel classification in order to match the cartographic characteristics of other data layers or to reduce the number of vertices and polygons. Though rules have been set up to convert raster data to vectors, the effects of generalization in such operations has not been rigorously explored. Methods of quantifying the spatial variation between vector-to-raster and raster-to-vector conversions must be developed.

The working group also identified a need for additional information on remote sensing positional error characteristics and the correlation between positional and classification errors. More knowledge is required on the characteristics of alternative remote sensing platforms and how advances in global positioning system (GPS) technology will improve remote sensing data positional accuracy. The incorporation of elevation correction in georeferencing procedures for both aircraft and satellite remote sensing data is critical to achieve acceptable positional accuracy for incorporation into a GIS. Research needs to be conducted to assess the relationship of elevation model scale and the degree of relief displacement in the georeferencing process. Furthermore, guidelines and procedures need to be established for geodetic control requirements in the georeferencing process, including appropriate datum selection.

Because the goal of an integrated remote sensing and GIS analysis is generally the presentation of information to a decision maker, derived products must be designed to communicate information regarding the accuracy of information products as well as the thematic content. The working group identified a need to develop geometric and thematic reliability diagrams for remote sensing and GIS results that convey the spatial characteristics of uncertainty in information products.

1.4. Data structures and access to data

Ehlers et al. (1991) describe the findings of the data structures and data access working group, and findings of their paper are summarized in this section. A number of research issues still exist with regard to technical choices of the data structures used to encode and transfer spatial data and information. Conversions between spatial representations are not always error free, and the nature of the errors are not well understood, nor are they well documented. Furthermore, these systems generally do not provide a prospective user with the tools and information necessary to evaluate the suitability of the data for the user's needs. The Data Structures and Access group divided key research issues into four subtopics:

(1) Concepts of Space in GIS and Remote Sensing,
(2) Data Conversion and Exchange,
(3) Integrated Databases, and
(4) Display of User Interfaces.

Remote sensing employs a field-based model of space that is systematically sampled by the sensing system. GIS may be built from this perspective or from an object-oriented point of view. There are substantive differences in the levels of abstraction and precision between these representations. The working group identified a need to determine a unifying theory or model that would permit one to use the potentially different representations of remote sensing and a GIS to produce an integrated model of space. This would potentially allow one to define the optimum transformations between different representations of geographic space. If lossless transformations of spatial information are not possible, then we must determine if the information loss can be quantified.

The working group identified a number of unanswered research questions about the conversion and exchange of spatial data. For example, what characteristics of remote sensing data are required for various GIS applications? The group identified a need to determine which data structures and data management strategies are optimal for GIS-guided image interpretation. Optimal exchange/conversion methods for enhanced access to large, distributed cartographic and remote sensing databases need to be developed. The group also identified the potential need for integrated test data sets containing image, cartographic, field data, and model results to test these issues.

Participants found that a key to better remote sensing/GIS integration may be the construction of database management systems (DBMSs) that provide services specific to spatial data handling. Key research issues in this area would include determining the degree and nature of object orientation within such a system and whether a monolithic or modular DBMS architecture would be optimal. The inability of existing spatial DBMSs to handle different data models in an efficient and transparent manner was identified as an impediment to further integration of remote sensing and GIS technologies. Can consistent terms for describing features be developed between GIS and remote sensing?

With respect to data access, participants felt that current inventory mechanisms that provide descriptions of, and access to, existing GIS and remote sensing data are inadequate. This is particularly true as we look to future, advanced satellite sensor systems. Research topics that must be addressed to remove existing limits to integration include the further development of spatial data exchange standards, the specification of minimum standards for the digitizing and encoding of spatial data, and the identification of existing datasets that would prove of greatest value to a wide array of users.

1.5. Data processing flows and methodology

Discussions regarding the topic of data processing flows and methodology were presented by Davis et al. (1991) and are summarized here. This working group focused on basic scientific issues that must be addressed to improve our understanding and characterization of systems using remote sensing and a GIS. The research community has been shifting from empirically based image classification, mapping, and inventory to more deterministic modeling of scene characteristics based on the physical laws of radiative transfer and energy balance. Similarly, GIS analyses are increasingly sophisticated, moving from simple map overlay and relational models to spatially distributed simulation modeling. Continuing progress in this area is dependent upon better integration of remote sensing and GIS technologies. As remote sensing and GIS methods evolve and full integration becomes even more of a reality,

analysis will be increasingly limited by our understanding of phenomena and
their representation in spatial databases. Participants of this working group
identified four subtopics for intensive consideration. These were:

(1) Scaling of Geographic Phenomena,

(2) Measurement and Sampling of Geographic Variation,

(3) Monitoring and Change Detection, and

(4) Data Processing and Information Flow.

   The working group suggested the need for improved mathematical and statistical approaches to document the accuracy and scale dependence of data
products. This would require a better understanding of the ways in which
measurement strategies influence accuracy and scale dependence. The group
also identified a need to determine optimal measurement and analysis methods to model specific processes and to determine the changes in information
content and accuracy associated with transformations in the data processing
flow. Transformations may create fundamental changes in the nature of data,
which may render them inappropriate for certain analyses. To address this
problem it was recommended that the ability to propagate spatial attribute information of data layers be more fully developed in data processing systems.
This would help to define the required characteristics of a lineage tracking
system, which would allow posterior analysis of assumptions and methods
used in the data processing flow for a particular information product.

   Impediments to the integration of remote sensing with more conventional
spatial data were identified as being conceptual as well as technical. The
working group identified a need to understand integrated processing methods
within the larger framework of geographical analysis, encompassing sampling, cartography, remote sensing, and GIS. This would involve improving
understanding and methods for measurement, sampling, simulation modeling, and validation, as well as developing ways in which the synergistic nature
of these methods may be captured. Problems identified by the working group
include defining appropriate strategies for data acquisition and spatial modeling, adequate characterization of scale dependence in surface features, and
the relationships of absolute versus relative scale within a remote sensing
context. Improved methods are needed for measuring spatial properties of
data, such as spatial autocorrelation, two-dimensional spectral analysis, and
block variance analysis. What other analytical tools should be included in
an integrated geographic information system and what statistical approaches
are appropriate for analyzing IGIS products? Among these developments,
the group identified a need for software that facilitates sensitivity analysis of
complex spatial models and the visualization of IGIS products.

1.6. Future computing environments

The rate of change of computing technology and computing capabilities continues to increase. A paradigm of networked computing facilities has clearly evolved in the spatial data processing community. This paradigm also has accompanying requirements for distributed data management and computing. Operations that once took hours or days on mainframes can now be performed in real time, near-real time, or real-enough time on microcomputers and workstations. In the face of these advances there is a clear need to examine the ways in which future computing environments may improve the integration of remote sensing and GIS technologies. This section summarizes findings of the Future Computing Environments working group, which were documented by Faust, Anderson, and Star (1991). The reader should be aware that the rapid evolution of the computer industry makes this area of investigation most sensitive to obsolescence. Still, many of the ideas developed in this working group have not yet been thoroughly tested or implemented. Three general areas of emphasis were identified by the Future Computing Environments group, including:

(1) Visualization,
(2) Parallel Processing, and
(3) Software Environments.

The working group identified areas where research into developing computer technologies could reduce barriers to human analysis in an integrated GIS environment. Access to remote sensing and GIS technologies could be facilitated for nonspecialist users by employing expert systems in the user interface design. The barrier between analyst and data may be further reduced by employing multimedia approaches to develop an "optimum working environment." Visualization will be an important component of this optimum working environment. Visualization methods that provide a clear understanding of multidimensional measurement space must be further developed and should probably include tests of the usefulness and applications of stereo displays. The vast amounts of data needed to understand the spatial and temporal dynamics of the integrated global environmental system make visualization techniques not only useful, but absolutely necessary.

A further area where computing research may reduce the impediments to remote sensing and GIS integration is the development of methods that increase the speed and thus the interactive nature of such analyses. The group identified the need to determine the possible benefits of parallelizing computational processes. This would include a determination of which parallel processing methods are appropriate in a given case. The potential of SIMD (single instruction, multiple data) and MIMD (multiple instruction, multiple data) methods should be explored for various analysis functions. In addition, the possibility of parallel computing in the analysis of vector data structures should be explored. The concept of parallel computing should also be generalized to include parallel analysis and tests of the practicality of remote sensing/GIS functionality across a heterogeneous network. There remains an incomplete understanding of the potentials and limitations of real-time, multi-investigator, spatial analysis.

Other research priorities highlighted by this group included software implementation techniques that would enhance the ability to deal with the enormous volumes of spatial data that will be used in integrated global analyses. One research area that was singled out was the expansion of data structures to handle information on the cataloging, processing history, and archiving requirements of various data products. The barrier of data volumes in an integrated analysis could also be addressed by testing the feasibility of dynamic compression and decompression techniques and developing standards for raster-to-vector conversion. Finally, the group suggested that the development of a standard set of low-level IGIS functions would help to diversify the roles of remote sensing and GIS by facilitating the integration of these methods into related software applications.

1.7. Institutional issues

From international to national and from regional to local levels, institutional issues influence the application of remote sensing and GIS. The development of these two technologies has been to a greater or lesser extent conditioned not only by the U.S. Federal establishment but, also governmental agencies, private industrial concerns, and academic institutions around the world. Indeed, the products of remote sensing in most countries are subject to strict governmental control. GIS products, although not currently subject to the same level of restriction, could in some circumstances come under the same types of control in the future. Technology acceptance, technology integration, issues of data sharing, and the value of data in policy and management decision making are all important concerns. This section summarizes the results of the Institutional Issues group and is taken from the paper by Lauer et al. (1991). This working group identified six areas of concern:
(1) Data Availability,
(2) Data Marketing and Costs,

(3) Equipment Availability and Costs,
(4) Standards and Practices,
(5) Education and Training, and
(6) Organizational Infrastructures.

An examination of institutional impediments to integrating remote sensing and GIS suggests a variety of opportunities for investigation. The group identified a need to evaluate how spatial data is used in decision making and how spatial data is managed in the public sector. The potential costs and benefits of various data sharing policies and mechanisms should be investigated. In addition to data availability, hardware and software availability can create impediments to integrated spatial data analysis. The working group suggested that investigations be made into the advantages and disadvantages of various computing environments. This would include consideration of potential problems with single- versus multiple-vendor environments, both within and between organizations.

The development and adoption of standards is a key institutional concern. Research was suggested to define common geoprocessing languages for interdisciplinary use. This may include the development of a glossary of common terminology and friendlier user interfaces. It would also be useful to explore methods by which standards can be most easily developed and best documented.

The working group identified the need to further develop training and educational materials for integration of remote sensing and GIS technologies and for implementing these technologies in public agencies. The possibility of using novel methods for the creation and distribution of these materials was suggested, including the use of advanced telecommunications and various media.

It was suggested that models for interagency collaborations be developed to remove institutional impediments. This approach would include the development of incentives for forming creative consortia among various levels of government, academia, industry, and professional organizations. Such a framework would require an assessment of the optimal organizational bodies for handling specific implementation issues. It can be argued that with the recent push to "reinvent" government and the budget reducing initiative of the 1995 Republican Congress, collaboration and restructuring are beginning to occur.

1.8. Conclusion

The issues raised by Initiative 12 become clearly significant when one considers that the governments of the world are spending billions of dollars preparing remote sensor systems to monitor our environment. Meanwhile, industry continues GIS developments in the applications arena. It is critical that a complete infrastructure exists to bring these capabilities together. Accurate and timely information are required on the status and trends of socioeconomic, biophysical, and geochemical processes in our environment. Remote sensing offers us the only practical means for the collection of many types of basic environmental data at any scale beyond the local. GIS techniques must be employed to integrate diverse spatial data sources, extract information, and convey that information to decision makers in an effective fashion. We have argued that remote sensing and GIS technologies are intrinsically linked. To be effective, key GIS data layers must have complete spatial coverage and be as up to date as practical. Remote sensing can provide this capability. Remote sensing data often require correlative data to improve the accuracy of derived data products. GIS data layers can supply such information. However, as identified by Initiative 12, serious impediments remain to the seamless integration of remotely sensed and GIS data.

The following seven chapters address some of the issues raised in the I-12 activity. The authors include both I-12 participants and other invited experts. Chapter 2 by Manfred Ehlers addresses issues and recent developments in image rectification and registration. Chapter 3 by John Jensen, Dave Cowen, Sunil Narumalani, and Joanne Halls focuses on the topic of change detection. This is followed by a chapter on visualization by Nickolas Faust and Jeffrey Star. A leading-edge integrated modeling environment named *Amazonia* is presented by Terence Smith, Jianwen Su, and Amitabh Saran in Chapter 5. A discussion of accuracy and uncertainty within an integrated processing flow is then presented by Kenneth McGwire and Michael Goodchild. At the local scale Timothy Foresman and Thomas Millet focus on remote sensing and GIS integration in planning in Chapter 7. A global perspective is then taken by Gassam Asrar in Chapter 8 on GIS requirements in global change research. Finally, John Estes and Jeff Star provide a conclusion that distills their impressions of the I-12 activities into a prioritized research agenda for integrated GIS and provides a look to the future.

The following chapters define pieces in the puzzle of an improved remote sensing and GIS implementation. The overall I-12 effort attempted to provide an image of what the future of integrated remote sensing and GIS analysis should look like, suggesting how the various pieces should fit together. Hopefully, this effort will spur coordinated efforts by public, private, and academic sectors to develop new analytical capabilities that will improve our understanding and decision making with respect to society and the environment.

2

Rectification and Registration

Manfred Ehlers

2.1. Datasets in IGIS

A fundamental requirement for integrated processing of remotely sensed data and GIS data is that they be spatially referenced. Only then can we be assured of not comparing "apples and oranges" if we have placed all information layers in a precise registration (Figure 2.1). This simple fact, however, is often ignored or downplayed in practice and can lead to unacceptable results (e.g., rivers that flow uphill, soybean fields on a highway, etc.). As a consequence of erroneous results based on geometric inaccuracies we may face strong resistance against the use of integrated systems. It is therefore necessary to be aware of potential error sources in the integration of remote sensing and GIS and to be able to identify the appropriate correction methods.

There are multiple sources for geometric distortions within GIS layers and remote sensing images. Due to the inherently different nature of the sources for remote sensing images and nonimage GIS data, we have to treat their respective errors independently. In an integrated GIS (IGIS), we must consider four predominant data sources: (i) cartographic data, (ii) socioeconomic (e.g., statistic) data, (iii) field data (points and transects), and (iv) image data.

Cartographic data: Most current GIS information is converted from maps that follow cartographic rules and specifications. Some of these specifications are well known, such as map scale, geometric fidelities, or map projections, and can be handled by well-known coordinate transformation methods (Maling, 1991). However, some map "errors," such as map generalization techniques, cannot be specified suitably for a computer program (Buttenfield and McMaster, 1991; Müller, 1991).

13

Socioeconomic data: This data type belongs to the class of nonspatial data and can be input into a GIS only with additional locational information (e.g., coordinates). As socioeconomic data are often aggregated, their geometric precision is usually rather low.

Field data: Field (or ground truth) data have always played an important role for validating accuracies in land use classification and mapping applications of remotely sensed imagery. These data are increasingly being collected in digital formats ("GIS-Pads") and with global positioning system (GPS)-controlled coordinate measurements. This allows for easy integration with geographic information systems.

Image data: Images from airborne and spaceborne platforms can be obtained in analog (e.g., photographic) or digital (e.g., scanned) formats. The spectral bandwidth of remotely sensed images ranges from the ultraviolet to the thermal infrared and microwave domain of the electromagnetic spectrum. Digital images are often seen as an inherent component of a raster-based GIS.

2.2. Integration of remote sensing data in GIS

These error sources have been discussed in several other research initiatives of the NCGIA and other institutions (Goodchild and Gopal, 1989; Chrisman,

Figure 2.1. All layers in an IGIS database have to be precisely registered to allow interpretive processing

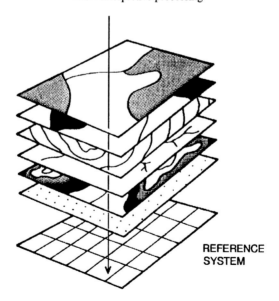

REFERENCE
SYSTEM

1991). The accuracy of integrated remote sensing and GIS analysis is also explored in Chapter 6 of this text. This chapter concentrates specifically on rectification and registration issues for integrated analysis of remote sensing and GIS data. Remotely sensed images can be differentiated according to their sensing principles and their recording platforms. Depending on their recording devices, we can differentiate four principle geometric mechanisms (Ehlers, 1992):

(i) Central perspective: The image is recorded in an extremely short time interval. Usually the recording device is a camera in which film is exposed (analog principle). Digital cameras employing two-dimensional arrays of charge-coupled devices (CCDs) are only slowly gaining ground owing to their limited resolution (current maximum in the range of 4,000 × 4,000 pixels). Desktop scanners are being used to convert analog photographic material into digital format. It has to be noted, however, that each transformation usually produces an information loss. To obtain all the information available from an aerial photograph we have to employ high-quality photogrammetric scanners that provide, among other specifications, a sampling rate of about 10 μm (or approximately 2500 dpi) (Ehlers, 1991).

(ii) Along-track (line, pushbroom) scanner: An along-track scanner records one line of an image per recording step with up to thousands of CCD sensors mounted in a linear array. The image is created as a series of lines resulting from the forward motion of the recording platform. An example for this technology is the High Resolution Visible (HRV) sensor on the *SPOT* satellites.

(iii) Cross-track (mirror, whiskbroom) scanner: Using a rotating or oscillating mirror, one single sensor can in principle create a whole image. The rotation of the mirror is utilized to record one line; the forward motion of the platform is again used for the second dimension. Examples include the *Landsat* Multispectral Scanner (MSS) and Thematic Mapper (TM) sensors.

(iv) Active sensors (radar, sonar): Active sensors transmit electromagnetic energy (or in the case of sonar, acoustic energy) and record the energy reflected back to the sensor. The time difference and intensity of the backscattered signal are used to create an image. For image formation, these sensors have to point sideways, which creates a number of specific geometric problems.

Platforms that have been utilized to record remotely sensed images are predominantly aircraft and satellites. In principle, both platforms exhibit the same geometric distortions. In practice, however, they can be treated differently. Satellite remote sensing has to account for Earth rotation and, in some instances, also the curvature of the Earth. On the other hand, their relatively undisturbed orbits give them the advantage of high stabilities in altitude, attitude, and velocity. Also, terrain relief is of very little influence on the recording process for most satellite sensors. This improvement in geometric accuracies, however, is usually at the expense of resolution.

To obtain high-resolution imagery, it is in most cases necessary to use airborne remote sensing devices. Aircraft are usually not affected by Earth parameters such as curvature and rotation during the recording process. They are, however, considerably more prone to distortions in their flight path (altitude and speed variations, pitch, roll, and yaw) and panoramic effects related to the imaging geometry (Richards, 1986) (Figure 2.2).

Other geometric distortions are related to the recording sensor (e.g., scan rate, mirror nonlinearities in cross-track scanners, misalignment in along-track scanners, sensor wide field of view, or side-looking geometries in active scanners).

2.3. Fundamentals for rectification and registration

Before the images can be integrated with a GIS (or "simply" overlaid on a map), the geometric distortions have to be corrected. Before we can discuss techniques and algorithms for geometric corrections, it is necessary to define a glossary of terms that we often find confused or incorrectly used.

Referencing: Referencing is the techniques that reference image pixels to a "master" coordinate set. This set can be pixel coordinates of a reference (master) image (relative referencing) or a geographic/geodetic coordinate system (absolute referencing or georeferencing). This process does not involve an actual geometric transformation of the image to be referenced (often referred to as a "slave" image) but rather the calculation of a coordinate transformation function.

Registration: Registration is the process that comes after the computation of a coordinate transformation function. It involves the actual pixelwise geometric transformation of the slave image into the geometry of a master image or master dataset (e.g., GIS). We can then say that the image is "registered to a dataset." After this process, the image can be overlaid on a per-pixel basis. Registration also involves reformatting of output pixel sizes and assignment of new gray values.

Figure 2.2. Effect of platform position and attitude errors on the image
formation (from Richards, 1986)

Sources of Geometric Distortion

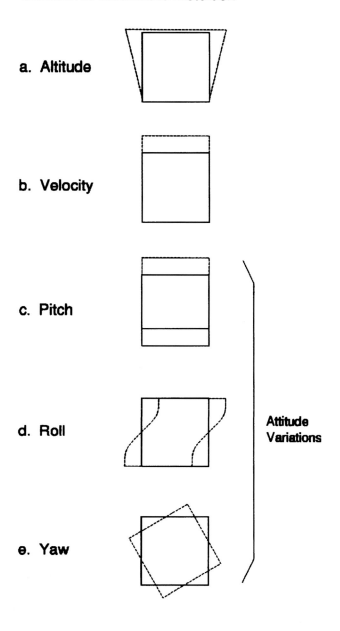

Rectification: Rectification is the process of registering a remote sensing image to a geographic/geodetic coordinate system (e.g., Universal Transverse Mercator or latitude/longitude coordinates). This process uses the same techniques as the registration process above, but now the master dataset is an absolute coordinate system. It is also called "geocoding." Registration/rectification techniques can be combined if an image is registered to a geocoded master image.

Resampling: Resampling is a part of the registration/rectification process. It involves the assignment of new gray values or digital numbers (DNs) for each pixel. As the location of registered or geocoded pixels will not usually project to an exact pixel location in the new image, new gray values have to be interpolated. Because this step involves a new sampling of data, it is called resampling.

In general, the process of integration of remote sensing with GIS or cartographic databases involves the following steps: (i) referencing; (ii) registration/rectification and resampling, and (iii) overlay and integration with the GIS database.

2.4. Geometric correction algorithms

2.4.1. Deterministic modeling

There are essentially two different techniques that can be used to correct the geometric distortions present in a remotely sensed image. One is the deterministic approach, where one models type and magnitude of the distortions and uses these models to correct the data. These techniques work well when the types of distortions are well known (e.g., Earth rotation, Earth curvature, or sensor characteristics). Efforts to model the actual flight path, however, have so far been more or less experimental because these models require extremely precise ancillary information about the platform's position and attitude values. Even for relatively stable satellite platforms, these models require more information than is available for the current generation of remote sensing satellites such as *Landsat*, *SPOT*, and *ERS* platforms.

More complicated techniques and more complete information are required to model aircraft flight paths. Theoretical solutions were already presented in the 1970s (see, for example, Baker, Marks, and Mikhail, 1975; Konecny, 1976). For practical purposes, however, flight parameters could not be obtained with the required accuracies. With the improvement of navigation systems and the advent of GPS, however, the first promising results have

Table 2.1. *Displacement in meters in SPOT imagery as a function of terrain of relief and camera viewing angle*

Viewing Angle	Terrain relief (m)				
	0	100	200	500	1000
0	0	0	0	0	0
5	0	10	20	49	98
10	0	20	40	100	200
15	0	31	62	153	305
20	0	42	84	209	418
25	0	54	108	271	542

been reported from experimental aircraft missions (Egels, 1991; Jacob, 1991; Fisher, 1992).

Yet even with integrated precise navigation parameters and accurate GPS positioning values, aircraft imagery will usually require knowledge of terrain features for a precise rectification/registration. This fact holds true for side-looking satellite sensors such as the European radar satellite *ERS-1* and for the *SPOT* sensors, which also provide off-nadir viewing capabilities from a camera that can be tilted $\pm 27°$ off-axis. To demonstrate the effects of terrain on the pixel locations, Table 2.1 presents the displacements for *SPOT* imagery calculated as a function of terrain and camera viewing angle. It can be seen that even a terrain with a relief of only about 200 m already yields a displacement of 3 pixels for a $10°$ viewing angle. With *Landsat* TM data, however, these effects can usually be neglected because of the near-nadir viewing geometry of the system.

It might prove possible in the future that this terrain information (e.g., digital elevation models) can be obtained from the imagery itself if stereo information can be provided (such is the case with *SPOT* or the German experimental scanner Stereo MOMS) or if it can be generated by other means, for example, through laser profiling or simply from an existing geographic information system. To date, however, these efforts have been largely experimental and have not been applied to satellite missions. For operational rectification/registration techniques, we rely on mathematical-statistical (stochastic) methods.

2.4.2. *Statistical modeling*

This approach depends on establishing an assumed mathematical relationship between the pixel locations in the remote sensing images and the pixel locations in the master image or the coordinates of those points on the ground, respectively. These models do not depend on any physical relationship as is the case with deterministic models. Because the parameters of the mathematical functions have been derived through standard statistical procedures, we may call these techniques statistical modeling.

The relationship between master and slave datasets can be used to correct the image geometry irrespective of the nature of the distortion. Although different mathematical functions may be used, the principle method applies to all of them. We assume that there exists a mapping function F, so that the pixel coordinates (x, y) of a slave dataset can be mapped into the coordinate system (X, Y) of a master dataset:

$$F(x, y) = (X, Y). \tag{2.1}$$

This mathematical relationship holds true for all pixels (x, y). If this function F is known, then we can calculate the exact location for each pixel and create a registered or rectified dataset. There are some problems, however, with this straightforward and elegant procedure. First, we do not know the explicit form of the mathematical mapping function. Second, we do not even know if such a function exists at all: Each individual pixel might require its own distinct mapping function (Konecny, 1976).

2.4.2.1. Polynomial mapping functions

In practice, these mapping functions were generally chosen as simple polynomials of first through fifth degree (see, e.g., Billingsley et al., 1983; Richards, 1986). The general form of a polynomial mapping function $(X, Y) = F(x, y)$ of nth degree is given in the following equation:

$$X = \sum_{i=0}^{n} \sum_{j=0}^{i} c_{ij} x^{i-j} y^{j}$$

$$\tag{2.2}$$

$$Y = \sum_{i=0}^{n} \sum_{j=0}^{i} d_{ij} x^{i-j} y^{j},$$

or explicit for a polynomial of second degree:

$$X = c_{00} + c_{10}x + c_{11}y + c_{20}x^2 + c_{21}xy + c_{22}y^2$$

$$\tag{2.3}$$

$$Y = d_{00} + d_{10}x + d_{11}y + d_{20}x^2 + d_{21}xy + d_{22}y^2.$$

If the parameters c_{ij} and d_{ij} are known, then the mapping polynomials can be used to relate any pixel in the slave dataset to a corresponding location in the master dataset. Values for these parameters can be estimated by choosing a set of features that can be identified in both datasets. These features are often referred to as control points (CPs) or – in the case of a geodetic or geographic reference coordinate system – as ground control points (GCPs). They have to be well defined and unambiguous and could be road or airport runway intersections, bridges over rivers, prominent coastline features, small islands, or the like. The proper procedures for selecting appropriate CPs is thoroughly discussed in Welch, Jordan, and Ehlers (1985).

By substituting the CP coordinates into the above equations we obtain a set of equations for the unknown polynomial coefficients. If we have enough CPs we can then solve this set of equations and calculate the unknown parameters. The necessary number of control points depend on the degree of the polynomial function. A polynomial of first degree, for example, requires only three control points, whereas polynomials of second, third, fourth, and fifth degrees require six, ten, fifteen, and twenty-one CPs. In practice, however, significantly more than the required minimum are selected and the coefficients are estimated using least squares techniques. With this, we will no longer have a perfect fit of our control points and can calculate the positional error in our registration/rectification procedure.

Nevertheless, one has to be aware that higher order polynomials will always yield less positional errors for the used CPs. This is just a statistical result of the fewer degrees of freedom in the parameter estimation process (Sachs, 1974). A more reliable way of calculating the residual positional error after the rectification is to get an independent estimate using withheld test or check points (Figure 2.3). These points are not used to calculate the polynomial coefficients. Using this method, Welch et al. (1985) could demonstrate that *Landsat* TM data were more accurately rectified when a polynomial of first order was used, although the residual errors at the CPs were lower with higher order polynomials.

2.4.2.2. Discontinuous interpolation functions

Whereas global polynomial modeling of image registration/rectification geometry usually shows good results for satellite imagery, there are a number of shortcomings when this method is applied to aircraft data. The disturbances of aircraft imagery (see Section 2.2) are often too severe to be compensated by global polynomials even of high order. Aerial photography can usually be easily corrected using the well-known photogrammetric equations for the camera imaging model (collinearity equations) (Konecny and Lehmann, 1984). Analog (hardcopy) photogrammetry requires using sophisticated photogrammetric equipment (analytical plotters) for this correction. With the advent of powerful and affordable scanner technology, however, it is possible to use photogrammetric restitution techniques in a completely digital environment (Ehlers, 1991; Steiner, 1991). It is quite evident that digital (softcopy) photogrammetry will play a major role in data acquisition and updating for IGIS (Ebner, Fritsch, and Heipke, 1991; Thorpe, 1991).

Image data from aircraft scanners, however, cannot be treated with such a rigorous model. Principally, every single line or even point has to be treated individually. A number of geometric restitution algorithms have been proposed in the 1970s and early 1980s. These range from Fourier series approximation to interpolation in Gauss-Markov fields and to finite element methods (see, e.g., Konecny, 1976; Ebner, 1979; or Schuhr, 1982).

Unfortunately, in practice, most of these methods were far too complicated and computer intensive to be used in operational systems. Most often, the standard polynomial approach was modified to a piecewise polynomial procedure. The image (or flight path) is broken up into a number of subimages and each is processed independently. Ehlers (1984a) used this method to successfully merge *Landsat* MSS and *SEASAT* radar satellite data. Jensen et al. (1987) showed the effectiveness of this approach for aircraft satellite data. A problem with this method, however, is the transition from one subimage to another because both subimages are treated as individual images. Consequently, some "massaging" is usually required for the transition zones, and polynomials of first order have to be used to avoid complications in the extrapolated parts of the images.

Figure 2.3. A graphical representation of registration/rectification using polynomials developed from CPs (shown as triangles) to transform the image data (left) to a reference base (right). Error vectors at test points (circles) after the rectification are used to calculate the RMSE$_{xy}$

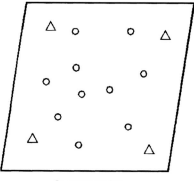
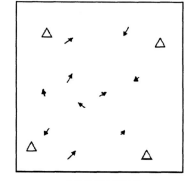

SLAVE IMAGE   **MASTER IMAGE**

A basic critique of this method is that we no longer have a general mathematical model that relates the slave image to the master dataset. Instead, we now have a number of mathematical relations that work independently in different parts of the datasets. These different parts are also being chosen on a random or nonspecified basis. Nonetheless, it is evident that when global polynomial interpolation functions do not work with sufficient accuracy, piecewise polynomial procedures (or any procedure that uses interpolation functions with discontinuities) can yield a significantly improved geometric registration accuracy. Yet even these discontinuous interpolation functions cannot cope with image datasets of high locally varying geometric distortions. In these cases the multiquadric interpolation method is a promising alternative.

2.4.2.3. Multiquadric interpolation function

The multiquadric function was first proposed by Hardy (1971) for the interpolation of irregular surfaces. This method was also actually a numerical type of biharmonic analysis and later renamed to multiquadric-biharmonic method (Hardy, 1990). It was modified by Göpfert (1982) and Ehlers (1987) to be used for image correction of remotely sensed data. In this form, it is particularly suited for the rectification of remote sensing images of large scale and locally varying geometric distortions.

The first step before the multiquadric algorithm is applied to an image is to use a standard polynomial rectification/registration routine. Polynomials of first or second degree should be used to keep distortions that can be introduced by polynomials of higher order to a minimum. This process keeps the remaining residuals small enough to be handled easier in a computer.

Within the coordinate system of the prerectified image we have a set of control point coordinates $(X_1, Y_2), \ldots, (X_n, Y_n)$, with (X_k, Y_k) as the (known) coordinates in the reference (master) system (e.g., UTM coordinates). As explained above, a polynomial rectification will usually yield residual errors at the used CPs. Let (X'_k, Y'_k), $k = 1, \ldots, n$, be the interpolated CP coordinates. Then the residuals (dX_k, dY_k) can be easily calculated:

$$dX_k = X_k - X'_k$$
$$dY_k = Y_k - Y'_k. \tag{2.4}$$

We can now "attach" these geometric improvements to all CP coordinates (X'_k, Y'_k) in the prerectified image to obtain the correct values (X_k, Y_k). The multiquadric method interpolates vectors (dx, dy) for every pixel (x', y') based on the CP improvements in the prerectified image. This is done in such a way that we model exactly the residual vectors (dX_k, dY_k) for each

CP. The values for all other pixels are interpolated using the distances to the CPs as weighting functions. With this method, we get a perfect fit for all CPs. We have to be aware, however, that this also means that we cannot check the quality of the final rectification at the CPs themselves because their residuals are zero.

The multiquadric procedure works as follows:

Step 1: Calculate the distance $f_j(x', y')$ between a point (x', y') and the CP (X_j, Y_j):

$$f_j(x', y') = [(x' - X_j)^2 + (y' - Y_j)^2]^{0.5}. \tag{2.5}$$

Step 2: Calculate the distance f_{ij} from CP (X_i, Y_i) to CP (X_j, Y_j):

$$f_{ij} = [(X_i - X_j)^2 + (Y_i - Y_j)^2]^{0.5}. \tag{2.6}$$

Note that $f_{ij} = f_{ji} = f_j(X_i, Y_i) = f_i(X_j, Y_j)$.

Step 3: Set up the interpolation matrix $F = (f_{ij})(n, n)$. [The (n, n) denotes that F is an n by n matrix.]

Step 4: According to Equation (2.4), we model our residual vectors DX and DY so that they can be calculated from F:

$$DX = FA \qquad \text{and} \qquad DY = FB$$

or explicitly,

$$
\begin{matrix}
dX_1 & & f_{11} \cdots & \cdots f_{1n} & a_1 \\
dX_2 & & f_{21} \cdots & \cdots f_{2n} & a_2 \\
\vdots & = & \vdots & \vdots & \vdots \\
dX_n & & f_{n1} \cdots & \cdots f_{nn} & a_n
\end{matrix}
$$

We get n equations for n unknowns in each set and can solve them for A:

$$
\begin{aligned}
f_{11}a_1 + f_{12}a_2 + \cdots + f_{1n}a_n &= dX_1 \\
f_{21}a_1 + f_{22}a_2 + \cdots + f_{2n}a_n &= dX_2 \\
\vdots \qquad \vdots \qquad\quad \vdots \qquad \vdots & \\
f_{n1}a_1 + f_{n2}a_2 + \cdots + f_{nn}a_n &= dX_n
\end{aligned}
\tag{2.7}
$$

This equation system has zeros at the diagonal and a symmetric matrix F. We now have to solve the equation system for A (or B, respectively) and can now model our residual improvements dX_k for $k = 1, 2, \ldots, n$:

$$f_{k1}a_1 + f_{k2}a_2 + \cdots + f_{kn}a_n = dX_k. \tag{2.8}$$

Step 5: We do the same with the Y-coordinates and vector B:

$$
\begin{array}{c}
dY_1 \\
dY_2 \\
\vdots \\
dY_n
\end{array}
=
\begin{array}{c}
f_{11} \cdots \cdots f_{1n} \\
f_{21} \cdots \cdots f_{2n} \\
\vdots \qquad \vdots \\
f_{n1} \cdots \cdots f_{nn}
\end{array}
\quad
\begin{array}{c}
b_1 \\
b_2 \\
\vdots \\
b_n
\end{array}
.
$$

We can now model our residual improvements dY_k for $k = 1, 2, \ldots, n$:

$$f_{k1}b_1 + f_{k2}b_2 + \cdots + f_{kn}b_n = dY_k. \tag{2.9}$$

Step 6: Our assumption is now that we can perform a geometric interpolation according to Equations (2.8) and (2.9) for every pixel (x', y') using the interpolation function $f_j(x', y')$ from Equation (2.5). Let f_j now stand for $f_j(x', y')$, with $j = 1, 2, \ldots, n$. Equations (2.8) and (2.9) will become:

$$f_1 a_1 + f_2 a_2 + \cdots + f_n a_n = dx$$

and $\hspace{10cm} (2.10)$

$$f_1 b_1 + f_2 b_2 + \cdots + f_n b_n = dy.$$

We have now the improvement vectors (dx, dy) to calculate the "true" location of each point (x', y'):

$$(X, Y) = (x', y') + (dx, dy). \tag{2.11}$$

If point (X, Y) is a CP with coordinates (X_k, Y_k), we get a perfect fit [see Equations (2.5) and (2.6) and keep in mind that $f_{ij} = f_{ji} = f_j(X_i, Y_i) = f_i(X_j, Y_j)$]. For all other points, we interpolate according to the multiquadric interpolation model above. The interpolation coefficients f_j provide a distance weighting function. The great advantages of the multiquadric algorithm are that (i) it describes a continuous interpolation function, (ii) all CPs contribute to the geometric transformation, and (iii) we can warp the image geometry in any given constraint (Figure 2.4). However, we cannot check the accuracy of this procedure at the control points as we are used to doing with polynomial mapping functions. We have a perfect fit for the control points. Consequently, we have to employ independent test points for accuracy checking.

2.4.3. Resampling

The actual assignments of new digital numbers or gray values to the calculated pixel coordinates is known as resampling. A resampled dataset is in the appropriate format to be integrated with other layers of an IGIS (e.g., overlay or image background). To create such a new data layer, we can principally make use of two different approaches: the *direct* (or *forward*) and the *indirect* (or *reverse*) *interpolation calculation*.

2.4.3.1. *Direct vs. indirect interpolation*

Within the direct calculation, the DNs of the dataset to be registered (input dataset) will be transferred on a pixel-by-pixel basis to the output dataset

Figure 2.4. Polynomial and multiquadric interpolation techniques. Control points are indicated as circles. Note the perfect fit at the control points and the continuous warping effects of the multiquadric algorithm

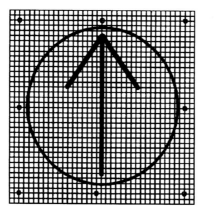

Control Points in Slave Dataset

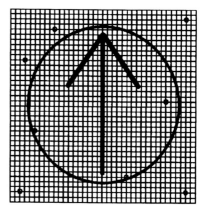

Control Points in Reference Dataset

Polynomial Interpolation (1st degree)

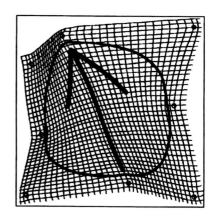

Multiquadric Interpolation

(Figure 2.5). For every input pixel the output location is calculated directly. Usually, the new pixel position is not an integer value and falls between several pixels of the output dataset. Consequently, the gray value of the input pixel has to be split between a number of output pixels (Figure 2.6). This process is also called *pixel carryover*.

This procedure, however, has a number of disadvantages: Output pixels can be addressed more than once, which poses high demands on storage capabilities and the coding algorithm. It is also possible that a number of output pixel locations will not be addressed at all. This leaves the output

Figure 2.5. Direct (forward) registration/rectification principle. For each pixel in the input (distorted) image, the location in the output image is calculated

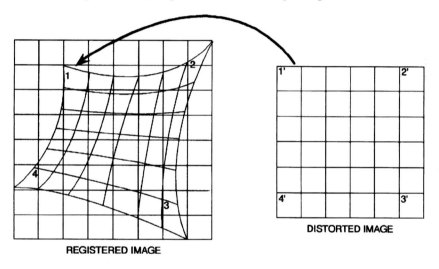

REGISTERED IMAGE

DISTORTED IMAGE

Figure 2.6. Pixel transfer within the direct registration/rectification (from Castleman, 1979)

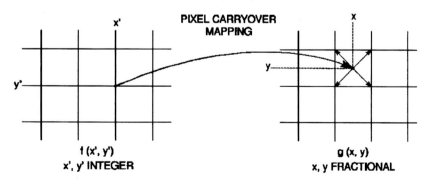

Ehlers

image with gaps that later on need to be interpolated. As a consequence, the indirect method is preferred for the integration of datasets.

The indirect method generates the output image on a pixel-by-pixel basis. For each output pixel the location of the mapped input pixel is calculated (Figure 2.7). Again, we cannot expect exact pixel matches, which means that the new value for each output pixel has to be calculated from the neighboring

Figure 2.7. Indirect (reverse) registration/rectification principle. The output image is calculated on a per pixel basis for registration

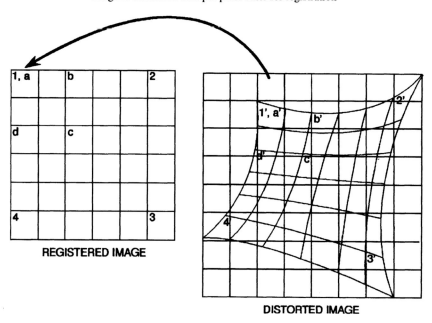

Figure 2.8. Pixel transfer within the indirect registration/rectification (from Castleman, 1979)

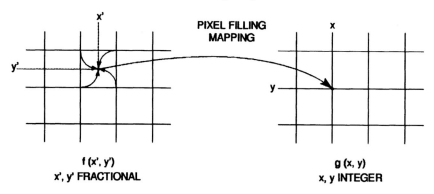

input pixels. This method is also referred to as *pixel filling* (Figure 2.8). With this procedure, every output pixel is addressed only once during the interpolation process. Also, gaps will not occur in the output datasets. In addition, if a registration procedure requires that other already registered datasets (e.g., digital elevation models) are to be used, they can be directly incorporated into this interpolation procedure.

2.4.3.2. DN interpolation

For the actual assignment of new DNs, a number of mathematical functions have been developed (see, e.g., Schowengerdt, Park, and Gray, 1984). As shown above, the indirect registration/rectification process calculates for each output pixel the respective position of its origin in the input image. This location is usually a fractional value and interpolation is necessary to determine the appropriate gray value (digital number) of the output image.

The simplest interpolation scheme is the nearest neighbor interpolation. In this case, the DN of the output pixel is taken from the input pixel nearest to the calculated location. This procedure is computationally simple and can be easily implemented. For remote sensing image data, however, it creates output images that are usually less appealing and often appear blocky. Especially for linear structures it introduces artifacts such as jagged edges. In a number of cases, however, it is imperative that the nearest neighbor interpolation technique be used. Data that cannot be interpolated (e.g., nominal values such as census data, soil classes, land use, and land cover classes) have to be resampled using nearest neighbor procedures. Also, if the original data await further processing (e.g., classification) it is preferable not to create new gray values but rather to keep the original ones.

The most often used interpolation technique in digital image processing is bilinear interpolation. This is done by fitting a hyperbolic paraboloid of the form

$$g(x, y) = ax + by + cxy + d \qquad (2.12)$$

to the DNs of four nearest pixels of the fractional input pixel location (Castleman, 1979). The four coefficients, a through d, are chosen so that g fits the known DNs at the four corners (Figure 2.9). To interpolate the value of $g(x, y)$ with $i < x < i+1$ and $j < y < j+1$, we perform a series of one-dimensional linear interpolations. First, we linearly interpolate in the x-direction between $g(i, j)$ and $g(i + 1, j)$ and between $g(i, j + 1)$ and $g(i + 1, j + 1)$:

$$g(x, j) = g(i, j) + x[g(i + 1, j) - g(i, j)]$$
$$g(x, j + 1) = g(i, j + 1) + x[g(i + 1, j + 1) - g(i, j + 1)]. \qquad (2.13)$$

Second, we linearly interpolate in the y-direction between the interpolated gray values:

$$g(x, y) = g(x, j) + y[g(x, j + 1) - g(x, j)]. \qquad (2.14)$$

Substituting (2.13) into (2.14) yields

$$g(x, y) = [g(i + 1, j) - g(i, j)]x + [g(i, j + 1) - g(i, j)]y$$
$$+ [g(i + 1, j + 1) + g(i, j) - g(i, j + 1)$$
$$- g(i + 1, j)]xy + g(i, j). \qquad (2.15)$$

Equation (2.15) is of the form of Equation (2.12) and is thus bilinear. Note that Equation (2.15) can also be derived by first performing two one-dimensional interpolations in the y-direction followed by one in the x-direction. Bilinear interpolation gives a much more visually appealing output result, especially for images, and is also not too computationally intensive. Consequently, it is the standard resampling algorithm for remotely sensed data if DN interpolation is permitted.

For certain geometric operations, especially when magnification is involved, the smoothing effects of bilinear interpolation may degrade some image detail. For these cases, higher order interpolation functions may be

Figure 2.9. Bilinear interpolation procedure. The new gray value is calculated using two horizontal (x-direction) and one vertical (y-direction) interpolations

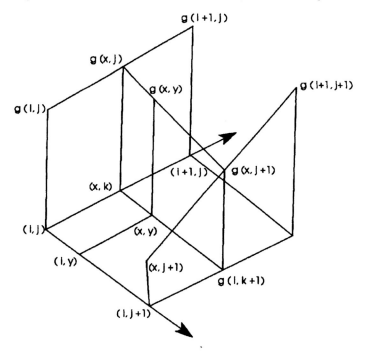

justified. Higher order interpolation functions involve more coefficients than Equation (2.12) and more than the four pixels. Examples include spline interpolation, Legendre centered functions, or $\sin(x)/x$ approximating functions (Castleman, 1979). Based on information theory, the $\sin(x)/x$ interpolation function is the optimal interpolation function (Ehlers, 1984b).

The most common way of approximating a $\sin(x)/x$ function is using a cubic convolution algorithm (Figure 2.10). The cubic convolution interpolation uses the sixteen surrounding pixels. Cubic polynomials are fitted along the four lines of four pixels surrounding the input pixel location (Figure 2.11). There are several cubic polynomials that are possible for the interpolation process (see, e.g., Schowengerdt, Park, and Gray, 1984). One of these algorithms was implemented by NASA as the resampling algorithm for *Landsat* TM data (Beyer, 1983):

$$
\begin{aligned}
g(x, j) = {} & g(i, j)[4 - 8(1 + dx) + 5(1 + dx)^2 \\
& - (1 + dx)^3] + g(i + 1, j)[1 - 2dx^2 + dx^3] \\
& + g(i + 2, j)[1 - 2(1 - dx)^2 + (1 - dx)^3] \\
& + g(i + 3, j)[4 - 8(2 - dx) + 5(2 - dx)^2 - (2 - dx)^3].
\end{aligned}
$$

$$(2.16)$$

Figure 2.10. $\sin(X)/X$ and the cubic convolution response curves

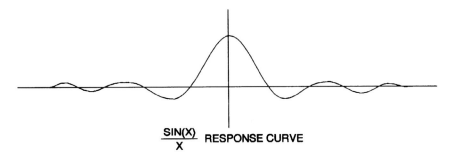

$$\frac{\text{SIN(X)}}{\text{X}}$$ RESPONSE CURVE

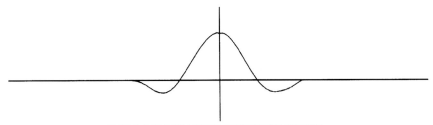

CUBIC CONVOLUTION RESPONSE CURVE

This expression is calculated for each of the four lines in the y-direction to yield the four values $g(x, j)$, $g(x, j+1)$, $g(x, j+2)$, and $g(x, j+3)$. These values are again interpolated using Equation (2.16) in the y-direction to give

$$
\begin{aligned}
g(x, y) = \; & g(x, j)[4 - 8(1 + dy) + 5(1 + dy)^2 - (1 + dy)^3] \\
& + g(x, j+1)[1 - 2dy^2 + dy^3] \\
& + g(x, j+2)[1 - 2(1 - dy)^2 + (1 - dy)^3] \\
& + g(x, j+3)[4 - 8(2 - dy) + 5(2 - dy)^2 - (2 - dy)^3].
\end{aligned}
$$

$$(2.17)$$

Figure 2.11. Cubic convolution interpolation is performed as four one-dimensional horizontal interpolations and one vertical interpolation

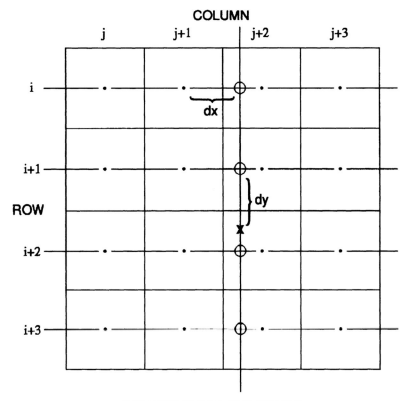

SQUARE CELLS ARE PIXELS

• PIXEL CENTER

x RESAMPLE LOCATION

○ HORIZONTAL INTERPOLATION POINTS

Cubic convolution interpolation yields an image product that is generally the most appealing in appearance. It is often used as output product for visual analysis or to serve as an image layer in an IGIS. Cubic convolution is generally only applied to image data. Similar to bilinear interpolation, it should not be used if subsequent digital image processing has to be performed with data where it is essential to keep the original gray values.

In many cases of integrating datasets it might not be necessary to actually perform a resampling. Data can be stored in their original format so that access to the unchanged data is always possible. If diverse datasets have to be integrated (e.g., change detection or cross-correlation analysis) they can be resampled "on the fly." This requires, of course, that there is enough computer power to provide such capabilities (Faust et al., 1991).

2.5. Two-dimensional vs. three-dimensional aspects

The knowledge of elevation and information derived from this knowledge (e.g., aspect or slope) is often essential for GIS applications (e.g., watershed management, road planning, etc.). Terrain information also plays a prominent role in the interpretation of remotely sensed data and their integration with GIS. The importance of terrain knowledge for the rectification/registration of airborne and satellite scanner data has already been stressed in Section 2.1. Although most current GIS provide some capabilities for handling the third dimension, this is usually done in the form of a digital elevation model as another layer within an IGIS. This means that elevation values are treated as attributes of two-dimensional (X, Y) coordinates. This makes most GIS at maximum a two-and-one-half dimensional (2.5D) system (see Chapter 4).

If elevation data have to be used in the rectification process, we have to use differential rectification algorithms (Konecny, 1979). One possible method to achieve this is to employ the well-known photogrammetric collinearity equations of the form of Equation (2.18). They include the elevation z as an explicit parameter:

$$X = \frac{a_{11}(x - x_0) + a_{12}(y - y_0) + a_{13}(z - z_0)}{a_{31}(x - x_0) + a_{32}(y - y_0) + a_{33}(z - z_0)}$$

$$(2.18)$$

$$Y = \frac{a_{21}(x - x_0) + a_{22}(y - y_0) + a_{23}(z - z_0)}{a_{31}(x - x_0) + a_{32}(y - y_0) + a_{33}(z - z_0)}.$$

However, these equations have to be modified for nonphotogrammetric data and are computationally much more complex than the polynomial equations of the form (2.2). It is, unfortunately, not possible to extend the polynomial equations to include z-values as well. Polynomials represent global mapping function parameters. If associated with z-values they would not be able to model varying terrain, which, of course, is exactly the reason why we wish to employ terrain values in these equations. It might prove possible to modify the multiquadric interpolation technique to consider height differences. The multiquadric interpolation algorithm does indeed consider locally varying geometry and can thus be extended to incorporate z-values. Such a modification, however, has not been tested to date and is still in the development stage.

Another procedure to cope with terrain relief has been used by Welch (1987) and Ehlers (1987). By selecting control points at mid-level altitude they were able to rectify satellite images using global mapping polynomials. Digital elevation data were introduced to differentially rectify the satellite images. In a second step they could use a simple one-dimensional resampling algorithm (adapted to the imaging mode of the sensor) to create relief corrected images (ortho-images). This approach, however, would still not work with aircraft data. Here, we would have to revert to the other methods described above.

2.6. Error analysis

Error analysis in geographic information systems is one of the most critical issues in current GIS research. In combining layers of diverse information with diverse accuracies, we are already placed in the position where a coherent theory of errors is impossible to use. For example, if we do an environmental assessment for an optimal location of a dump site, we might have to combine GIS layers that are digitized from soil maps, cadastral information from a town office, topographic information from large-scale maps, elevation data from a digital elevation model, geologic information from field probes, information about critical biotopes from field surveys, and vegetation analysis from aerial photographs. All this is, without any doubts, spatial information and as such is suitable for integrated analysis in an IGIS. But all layers have different origins, have different acquisition specifications, and thus have different spatial errors. Their nature is inherently different.

An IGIS has to combine data that represent absolute values, such as digital elevation data, with datasets of ordinal values, such as soil data, and with nominal data, such as the biotope information, which might be ranked according to an ecologic scheme. A thorough analysis has to present figures or, at least, estimates of the accuracy of the derived results. If the combination of errors exceeds a certain limit, the analysis might not be worth anything at all.

This limiting value has to be stated; otherwise the IGIS will be blamed for the bad decision. It is therefore essential that remote sensing and GIS researchers develop a theory of error for IGIS analyses.

Although some advanced error theories for certain information layers in GIS, such as for positional errors of polygons, have been developed, there is no universal error theory for IGIS (see, e.g., Goodchild and Gopal, 1989; Chrisman, 1991; Thapa and Bossler, 1992). It came as no surprise that the first NCGIA research initiative dealt with GIS and accuracy. With the inclusion of remote sensing data in IGIS we have to consider even more error sources in our integration model (Lunetta et al., 1991). We might have to face the fact that there is no general error theory for IGIS, and efforts to derive some error analysis within the scope of integration of remote sensing and GIS will have to work on limited aspects only. It is, however, necessary that we develop some kind of error quantification mechanism so that we can attach accuracy values to the results of an analysis that combines remote sensing and GIS.

One of the most critical areas in the accuracy issue for integrating remote sensing and GIS is that of errors introduced during the registration/rectification process. These errors can at least be addressed in a statistical manner. A commonly used procedure to quantify the geometric error in the registration/rectification process is the calculation of residual vectors for the employed control points. However, because these points are used to calculate the geometric transformation parameters they cannot serve as independent indicators. Well-distributed test points that are not used to establish the geometric transformation coefficients yield a much more precise estimate of the residual registration/rectification error (see Figure 2.3).

Independent test points have to be employed for all methods that do not use control points (e.g., deterministic modeling procedures) or have zero residuals at the CPs (e.g., multiquadric interpolation). Often, a thorough analysis of control points can help to identify unreliable points or to detect gross errors before the actual computation takes place. As a rule of thumb, we might be able to say that the quality of the selected control points will be highly correlated with the quality of the final rectification/registration.

2.7. Outlook and future work

The quality of the rectification can also be enhanced if automated matching techniques can be applied (Ehlers, 1984a). For the registration of similar datasets, these techniques can match identical points with greater accuracy than can be achieved with visual analysis (Mikhail, Akey, and Mitchell, 1984).

These techniques usually still require some preprocessing or interaction to identify possible CPs in the reference image. The use of fully automated registration procedures (e.g., the selection of the proper geometric procedure, the identification and matching of CPs, and the error analysis) remains an active field of research. For datasets that are similar in their characteristics, some positive results could be achieved (see, e.g., Ton and Jain, 1989; Ventura, Rampini, and Schettini, 1990; Fuller and Ehlers, 1991).

This line of research for automated geometric processing and integration will have to be continued at an accelerating pace. We will not be able to integrate multiple remote sensor data in environmental databases and geographic information systems if we have to depend on manual processing. Integrating terabytes of data per day will make it impossible to measure control points in all these datasets. In addition, there are also unprocessed historical datasets (e.g., early aircraft and satellite scanner data, analog data that can be scanned and integrated into IGIS), which require geometric processing to be used for monitoring changes in our environment. It is therefore imperative to develop automated methods for integrating diverse datasets in geographic information systems.

3

Principles of change detection using digital remote sensor data[1]

John R. Jensen, Dave Cowen, Sunil Narumalani, and Joanne Halls

3.1. The nature of change detection

Information about change in the landscape provides valuable information on the processes at work. Therefore, it is not surprising that significant research has been undertaken to develop methods of obtaining change information (Dahl, 1990; Estes, 1992; Jensen, 1996). The information may be obtained by visiting sites on the ground and/or extracting it from remotely sensed data. Most change information eventually is placed in a geographic information system where it can be modeled with other data to derive new insight to the problem at hand. This chapter summarizes how change information is extracted from remotely sensed data. It first evaluates how remote sensor system and environmental parameters affect the change detection process. Some of the more important change detection algorithms are then identified and demonstrated. The chapter concludes with comments on the research needed in remote sensing/GIS integration for improved change detection.

3.2. Remote sensing system characteristics of importance when performing change detection

Failure to understand the impact of sensor system characteristics and environmental characteristics on the change detection process can lead to inaccurate results. It is useful to review these characteristics and identify why they can

---

[1]Some material in this chapter is used with permission from Dobson et al., 1995; Jensen, 1996; and Khorram et al., 1996.

have a significant impact on the success of a remote sensing change detection project.

3.2.1. Temporal resolution

Table 3.1 summarizes the temporal, spatial, spectral, and radiometric resolution of some of the most important satellite remote sensing systems used to obtain change information. There are two important sensor system temporal resolutions that should be held constant when performing change detection. First, it is best to use remotely sensed data acquired on anniversary dates (e.g., May 1, 1996 versus May 1, 1997). Using anniversary date imagery removes seasonal sun angle differences that can destroy a change detection project. Although it is usually not always possible to obtain exact anniversary dates because of orbit changes, cloud cover, etc., it is good practice to obtain dates as close as possible to one another. Second, the remote sensor data used should be obtained from a sensor system that acquires data at approximately the same time of day (e.g., *Landsat* TM data is acquired before 9:45 AM for most of the conterminous United States). This eliminates diurnal sun angle effects that can cause anomalous differences in the reflectance properties of the remotely sensed data.

3.2.2. Spatial resolution and look angle

At least two dates of remotely sensed data must be accurately registered to perform analytical change detection. Ideally, the instantaneous field of view (IFOV) of the sensor system is held constant on each date. For example, *SPOT* high resolution visible (HRV) multispectral data obtained at 20×20 meter spatial resolution (Table 3.1) on two dates are relatively easy to register to one another (Jensen et al., 1993b). Geometric rectification algorithms (Novak, 1992; Jensen, 1996) are used to register the images to a common map projection (e.g., Universal Transverse Mercator). The rectification process should result in the two images having a root mean square error (RMSE) of ≤ 1 pixel.

Sometimes it is necessary to use data collected from two different sensor systems in the change detection project, for example, *Landsat* MSS data (80×80 m) for date 1 and *SPOT* HRV data (20×20 m) for date 2 (Jensen et al., 1995a). Both datasets are then resampled to an acceptable minimum mapping unit pixel size (e.g., 20×20 m). Remember that the original IFOV dictates the information content of the remote sensor data. In this example,

Table 3.1. *Selected satellite remote sensing system characteristics*

Remote sensor system	Spectral resolution (μm)	Spatial resolution (m)	Temporal resolution (days)	Radiometric resolution (bits)
Landsat	Band 1 (0.50 − 0.60)	80 × 80	18	7[a]
MSS 1, 2 ,3	Band 2 (0.60 − 0.70)	80 × 80	18	7
	Band 3 (0.70 − 0.80)	80 × 80	18	7
	Band 4 (0.80 − 1.1)	80 × 80	18	7
Landsat Thematic	Band 1 (0.45 − 0.52)	30 × 30	16	8
Mapper 4, 5	Band 2 (0.52 − 0.60)	30 × 30	16	8
	Band 3 (0.63 − 0.69)	30 × 30	16	8
	Band 4 (0.76 − 0.90)	30 × 30	16	8
	Band 5 (1.55 − 1.75)	30 × 30	16	8
	Band 7 (2.08 − 2.35)	30 × 30	16	8
	Band 6 (10.4 − 12.5)	120 × 120	16	8
SPOT HRV XS	Band 1 (0.50 − 0.59)	20 × 20	pointable	8
	Band 2 (0.61 − 0.68)	20 × 20	pointable	8
	Band 3 (0.79 − 0.89)	20 × 20	pointable	8
SPOT HRV PAN	Pan (0.51 − 0.73)	10 × 10	pointable	8
AVHRR-12	Band 1(0.58 − 0.68)	1 × 1 km	daily	8
	Band 2 (0.72 − 1.10)	1 × 1 km	daily	8
	Band 3 (3.55 − 3.93)	1 × 1 km	daily	8
	Band 4 (10.3 − 11.3)	1 × 1 km	daily	8
	Band 5 (11.5 − 12.5)	1 × 1 km	daily	8
Indian	Band 1 (0.45 − 0.52)	72 × 72	22	8
IRS-LISS I	Band 2 (0.52 − 0.59)	72 × 72	22	8
	Band 3 (0.62 − 0.68)	72 × 72	22	8
	Band 4 (0.77 − 0.86)	72 × 72	22	8
Indian IRS-LISS II	same as above	36.25 × 36.25	22	8

[a]Landsat MSS 3, 4, and 5 collect data in 8 bits.

the 80 × 80 m MSS data resampled to 20 × 20 m pixels should not be expected to yield additional spatial detail.

The remote sensor data used in a digital change detection should be acquired with approximately the same look angle whenever possible. For example, some sensor systems like *SPOT* are pointable and collect data at look angles that are off-nadir by as much as ±20° (Table 3.1). If our goal was to inventory a cypress-tupelo wetland forest consisting of large randomly spaced trees, a *SPOT* image acquired at 0° off-nadir would look directly down upon the "top" of the canopy. Conversely, a *SPOT* image acquired at 20° off-nadir would record reflected radiant flux from the "side" of the canopy. Differences in recorded reflectance from the two datasets could cause change to be identified when in fact there was no change.

Table 3.2. *Land data satellites slated for launch between 1995 through 2004* (*Stoney, 1995*)

Country	Owner	Program	Launch date	Instrument type	Pan	M	Radar	#Color bands	Stereo type
Russia	G/O	Resours-02	1995	M		27		3	
China/Brazil	G/O	CBERS	1995	P&M	20	20		7	C/T
India	G/O	IRS-1 C	1995	P&M	10	20		4	C/T
Canada	G/O	RADARSAT	1995	Radar			9		
China/Brazil	G/O	CBERS	1996	P&M	20	20		7	C/T
Japan	G/O	ADEOS	1996	P&M	8	16		4	C/T
Russia	G/O	ALMAZ 2	1996	Radar			5		
United States	G/E	TRW Lewis	1996	P&M	5	30		384	
United States	G/E	CTA Clark	1996	P&M	3	15		3	F/A
United States	G/O	Earth Watch	1996	P&M	3	15		3	F/A
United States	G/O	Earth Watch	1997	P&M	1	4		4	F/A
France	G/O	SPOT 4	1997	P&M	10	20		4	C/T
United States	G/O	Orb View	1997	P&M	1&2	8		3	F/A
United States	G/O	Space Imaging	1997	P&M	1	4		4	F/A
ESA	G/O	ENVISAT	1998	Radar			30		
United States	G/O	Landsat 7	1998	P&M	15	30		7	
U.S./Japan	G/O	EOS AM-1	1998	M	15	15		14	F/A
Korea	G/O	KOMSAT	1998	P&M	10	10		3	F/A
United States	G/O	Space Imaging	1998	P&M	1	4		4	F/A
India	G/O	IRS-1 D	1999	P&M	10	20		4	C/T
France	G/O	SPOT 5A	1999	P&M	5	10		4	F/A
United States	G/O	EOS AM-2/L-8	2004	P&M	10	30		7	
France	G/O	SPOT 5B	2004	P&M	5	10		4	F/A

Notes: M = multispectral only, O = operational, F/A = fore & aft stereo
P = panchromatic only, G = government funded, C/T = cross, track stereo
P&M = pan and multispectral, C = commercially funded, E = experimental

3.2.3. *Spectral resolution*

The *number* and *dimension* of individual bands of remotely sensed data constitute the spectral resolution of a remote sensing system. Ideally, the spectral resolution is held constant when acquiring multiple-date imagery for a change detection project. When this is not possible, the analyst should select bands that approximate one another from the two different sensor systems (Jensen et al., 1995a). For example, *SPOT* bands 1 (green), 2 (red), and 3 (near-infrared) can be used successfully with *Landsat* MSS bands 1 (green), 2 (red), and 4 (near-infrared). Some change detection algorithms do not function well when bands from one sensor system do not match those of another sensor system, for example, utilizing the *Landsat* TM band 1 (blue) with either *SPOT* or *Landsat* MSS data is not wise.

3.2.4. Radiometric resolution

Radiometric resolution is defined as the sensitivity of individual remote sensing detectors to just-noticeable differences in the amount of reflected or emitted radiant flux from terrain of interest. Remote sensor data are normally collected at 8 bits with values ranging from 0 to 255 after analog-to-digital conversion (Table 3.1). Ideally, the remote sensor systems used to collect data for a change detection project need to have the same radiometric precision. When the radiometric resolution of data acquired by one system (e.g., *Landsat* MSS 1 with 7-bit data) are compared with data acquired by a higher radiometric resolution instrument (e.g., *Landsat* TM with 8-bit data), then the lower resolution data should be "decompressed" if possible to 8 bits for change detection purposes. It is important to remember, however, that the precision of decompressed brightness values can never be better than the original, uncompressed data. It is often necessary to radiometrically normalize the data obtained between dates using the normalization procedures to be discussed.

3.3. Environmental characteristics of importance when performing change detection

It is desirable to hold environmental variables as constant as possible when performing multiple date change detection.

3.3.1. Atmospheric conditions

Remotely sensed imagery collected on *n* dates to be used in a change detection project are rarely collected under identical atmospheric conditions. Subtle changes in humidity can have a deleterious effect while significant cloud cover on any date can be disastrous to the project. Therefore, scientists often try to use anniversary dates to ensure general, seasonal agreement between the atmospheric conditions on the *n* dates of imagery.

Many scientists attempt to remove the atmospheric attenuation in the imagery. Sophisticated atmospheric transmission models such as LOWTRAN (Kneizys et al., 1988) can be used to correct the remote sensor data if substantial in situ data are available on the day of the overflight. Alternative techniques include the use of empirical image normalization techniques to remove atmospheric effects (Eckhardt, Verdin, and Lyford, 1990; Hall et al., 1991; Jensen et al., 1995a). Image normalization reduces pixel brightness value (BV) variation caused by nonsurface factors so that variations in pixel

BVs between dates can be related to actual changes in surface conditions. Normalization enables the use of image analysis logic developed for a base year scene to be applied to the other normalized scenes. For example, consider *SPOT* data acquired on April 4, 1987 of Water Conservation Area 2A in the Florida Everglades that were radiometrically normalized to August 10, 1991 *SPOT* data (Figure 3.1). Pseudo invariant radiometric normalization targets that were assumed to be constant reflectors (e.g., nonturbid water bodies, constant bare soil areas, parking lots) were selected in a reference (base) image (1991) and nonbase dates of imagery (e.g., 1987). Any changes in the brightness values of the normalization targets were attributed to detector calibration, astronomic, atmospheric, and phase angle differences. Image normalization was achieved by applying regression equations to the nonbase year imagery that predicted what a given BV would be if it had been acquired under the same conditions as the reference scene (1991). The acceptance criteria for

Figure 3.1. (a) Rectified, normalized, and masked *SPOT* XS band 3 data obtained on April 4, 1987. (b) Rectified, normalized, and masked *SPOT* XS band 3 data obtained on August 10, 1991. (*Source*: Jensen et al., 1995)

Rectified, Normalized, and Masked SPOT Data of Water Conservation Area 2A

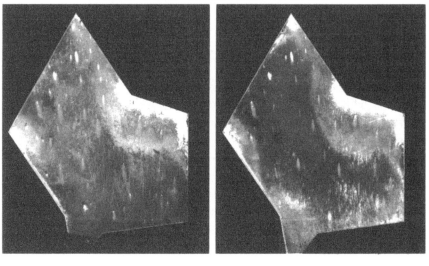

SPOT XS Band 3 4/4/87   SPOT XS Band 3 8/10/91

                                                        Meters
        10000   0   10000

(a)   (b)

potential "normalization targets" are summarized in Eckhardt et al. (1990) and Hall et al. (1991).

3.3.2. Soil moisture conditions

Soil moisture conditions should be held as constant as possible when selecting imagery for a remote sensing change detection project. It is important not only to look for anniversary dates that tend to hold seasonal soil moisture conditions constant, but also to review precipitation records to determine how much rain or snow fell in the days and weeks prior to the remote sensing data collection. When soil moisture differences between dates are significant for only parts of the study area (e.g., due to a local thunderstorm), it may be necessary to stratify (cut out) those affected areas and perform a separate analysis that can be added back in the final stages of the project.

3.3.3. Vegetation phenological cycle characteristics

When performing remote sensing change detection it is advisable to obtain near-anniversary images. This minimizes the effects of seasonal phenological differences that may cause spurious change to be detected in the imagery. One must also be careful about two other factors when dealing with man-made agricultural, rangeland, or forest landscapes. First, monoculture crops (e.g., red winter wheat) are not always planted at exactly the same time of year. A fifteen-day lag in planting date between fields having the same crop can yield serious change detection error. Second, the monoculture crops are not always of the same specie. Different species of the same crop can cause the crop to reflect energy differently on the multiple dates of anniversary imagery. Therefore, an analyst must know the biophysical characteristics of the vegetation as well as the cultural land-tenure practices of the study area. Most flawed remote sensing change detection studies are based on the use of imagery acquired for nonchange detection purposes. Remote sensing of change in vegetated landscapes requires careful selection of critical dates in the phenological cycle of the plants and a keen awareness of how the plants reflectance properties change through time.

3.4. Change detection algorithms

The selection of an appropriate change detection algorithm should be based on an analysis of: (1) the cultural and biophysical characteristics of the study area; (2) the precision with which the multiple-date imagery are registered; and (3) the utility, flexibility, and availability of change detection algorithms (Jensen, 1996). Five commonly used change detection algorithms are discussed:

- Post-Classification Comparison
  (Example: Florida Everglades – change in sawgrass and cattail.)
- Write Function Memory Insertion
  (Discussion: Par Pond, South Carolina – change in cattail and waterlily.)
- Image Arithmetic (Band Differencing or Band Ratioing)
  (Example: Par Pond, South Carolina – change in cattail and waterlily.)
- Manual, On-Screen Digitization of Change
  (Example: Sullivan's Island, South Carolina – Hurricane Hugo damage assessment.)
- Image Transformation (Multiple-Date Principal Components Analysis)
  (Discussion: General principles.)

Dobson et al., (1995), Khorram et al. (1996), and Jensen et al. (1996) provide some additional information on the application of these algorithms and identify other less commonly used change detection algorithms.

3.4.1. Post-classification comparison change detection

Post-classification comparison is the most commonly used method of quantitative change detection (Jensen et al., 1993a; Jensen, 1995a). It requires a complete classification of the individual dates of remotely sensed data (Rutchey and Velcheck, 1994). Unfortunately, every error in the individual date classification map will also be present in the final change detection map. Therefore, it is imperative that the individual classification maps used in the post-classification change detection method be as accurate as possible (Augenstein, Stone, and Hope, 1991; Price, Pyke, and Mendes, 1992).

To demonstrate the post-classification comparison change detection method, consider multiple-date classification maps of the Florida Everglades Water Conservation Area 2A shown in Figures 3.2a, b extracted from the normalized 1987 and 1991 *SPOT* data previously shown in Figure 3.1 (Jensen et al., 1995a). Five classes of land cover were inventoried on each date (shown in black and white). The 1987 and 1991 classification maps were compared on a pixel-by-pixel basis using an n by n GIS matrix algorithm whose logic is shown in Figure 3.3. This resulted in the creation of a *change image* (*map*), consisting of brightness values from 1 to 25. The analyst selected specific "from–to" classes for emphasis. Only a select number of the 20 ($n^2 - n$)

Figure 3.2. (a) Classification map of the Everglades Water Conservation Area 2A study area produced from April 4, 1987 *SPOT* XS data. (b) Classification map of the Everglades Water Conservation Area 2A study area produced from August 10, 1991 *SPOT* XS data. (c) Selected changes in land cover from 1987 to 1991 displayed using the logic found in Figure 3.3. (*Source*: Jensen et al., 1995)

Post-Classification Comparison Change Detection of Water Conservation Area 2A

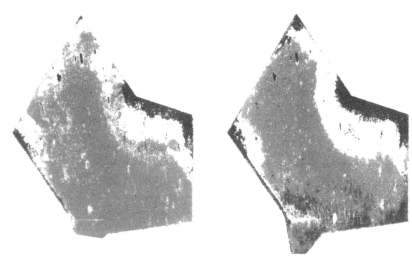

Classification Map of 4/4/87 SPOT Data

(a)

Classification Map of 8/10/91 SPOT Data

(b)

Change Map

Legend

- Brush
- Brush/cattail
- Cattail
- Sawgrass
- Sawgrass/cattail

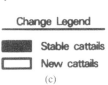

Change Legend

- Stable cattails
- New cattails

(c)

possible off-diagonal "from–to" land cover change classes summarized in
Figure 3.3 were selected to produce the change detection map (Figure 3.2c).
For example, all pixels that changed from any land cover in 1987 to "Cattail"
in 1991 were coded white (RGB = 255, 255, 255) by selecting the appropriate
"from–to" cells in the change detection matrix (3, 8, 13, 18, and 23). All pixels
that were cattails in 1987 and in 1991 (cell 13 in Figure 3.3) have a black look-
up table value (RGB = 0, 0, 0). If desired, the analyst could highlight very
specific changes such as all pixels that changed from "Sawgrass" to "Brush"
(cell 16 in the matrix) by assigning a unique look-up table value (not shown).

Figure 3.3. Basic elements of a post-classification comparison change detection
matrix used to produce the output change detection map shown in Figure 3.2c

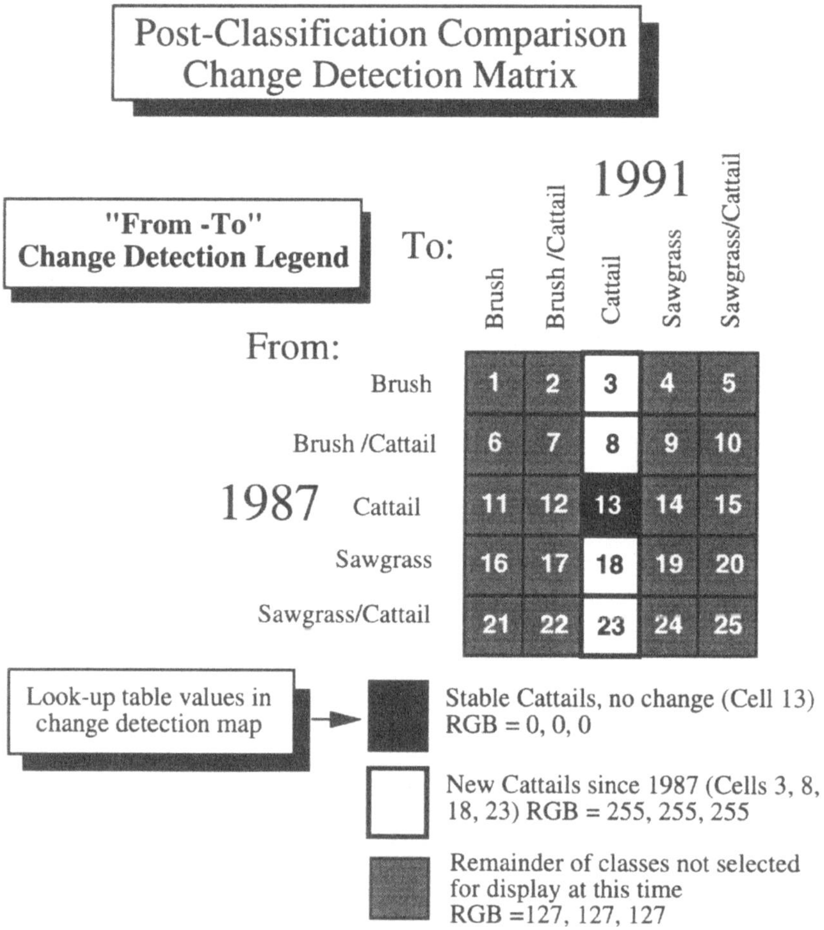

Figure 3.4. (a) Rectified and masked *SPOT* panchromatic data of Par Pond located on the Savannah River Site in South Carolina obtained on April 26, 1989. (b) Rectified and masked *SPOT* panchromatic data of Par Pond obtained on October 4, 1989. (c) A map depicting the change in waterlilies from April 26, 1989 to October 4, 1989 using image differencing logic. (*Source*: Jensen et al., 1993b)

Image Differencing Change Detection
of Par Pond in South Carolina

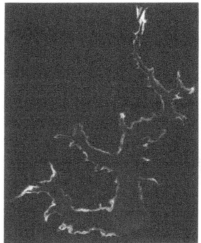

SPOT PAN April 26, 1989   SPOT PAN October 4, 1989

(a)   (b)

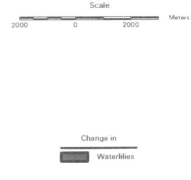

(c)

Post-classification change detection is easy to understand and implement (Jensen and Narumalani, 1992). Unfortunately, the accuracy of the change detection is dependent on the accuracy of the two separate classifications that are required. Khorram et al. (1996) provide some of the first innovative methods for assessing error in change detection maps produced using post-classification comparison change detection logic.

3.4.2. Write function memory insertion

Individual bands of remotely sensed data may be inserted into specific bands of image processing write function memory (red, green, and blue) to highlight change (Sader and Winne, 1992). For example, Jensen et al. (1993b) inserted *SPOT* panchromatic data of Par Pond in South Carolina obtained on April 16, 1989 into the green image plane and *SPOT* panchromatic data obtained on October 25, 1989 into the red image plane, and nothing in the blue image plane. The result was a dramatic display of the growth of aquatic macrophyte communities (cattail and waterlily) in Par Pond within a single year. Cattail were depicted in shades of yellow and waterlily in shades of red. In a separate analysis, they placed October 25, 1988 *SPOT* panchromatic data in the red image plane, October 4, 1989 data in the green image plane, and October 16, 1990 data in the blue image plane. The resultant image highlighted the change in aquatic macrophyte by year with stable areas being white. It is not possible here to demonstrate these methods in black and white.

3.4.3. Image arithmetic (band differencing or band ratioing)

Arithmetic operations may be applied to registered bands of imagery acquired on multiple dates to detect the change between images. Band differencing and ratioing are usually performed using a single band of imagery from each date. Image differencing involves subtracting the imagery of one date from that of another. The subtraction results in positive and negative values in areas of radiance change and zero values in areas of no change in a new "change image." In an 8-bit (2^8) analysis with pixel values ranging from 0 to 255, the potential range of difference values is -255 to 255. The results are normally transformed into positive values by adding a constant, c (e.g., 127). The operation is expressed mathematically as

$$\Delta_{ijk} = \text{BV}_{ijk}(1) - \text{BV}_{ijk}(2) + c,$$

where:

Δ_{ijk} = change pixel value

$BV_{ijk}(1)$ = brightness value at time 1

$BV_{ijk}(2)$ = brightness value at time 2

c = a constant (e.g., 127)

i = line number

j = column number

k = a single band (e.g., *SPOT* 3).

The change image produced using image differencing typically yields a Gaussian BV distribution, where pixels of no BV change are distributed around the mean and pixels of change are found in the tails of the distribution. Band ratioing involves exactly the same logic except that a ratio is computed and the pixels that did not change have a BV of 1 in the change image (Cablk et al., 1994). A critical element of both image differencing and band ratioing change detection is deciding where to place the threshold boundaries between "change" and "no-change" pixels displayed in the histogram of the change image. There are analytical methods that can be used to select the most appropriate thresholds in the tails of the distribution, such as the use of ± 2 standard deviations from the mean. However, most analysts prefer to experiment empirically, placing the threshold at various locations in the tails of the distribution until a realistic amount of change is encountered. The amount of change selected and eventually recoded for display is therefore subjective and must be based on familiarity with the study area.

Figure 3.4 depicts the result of performing image differencing on April 26, 1989 and October 4, 1989 *SPOT* panchromatic imagery of Par Pond in South Carolina (Jensen et al., 1993b; Jensen, 1996). The data were rectified, normalized, and masked using the methods previously described. The two files were then differenced and a change detection threshold was selected. The result was a change image showing the waterlilies that grew from April 26, 1989 to October 4, 1989 highlighted in gray (Figure 3.4c). The hectares of waterlily change are easily computed. Such information is used to evaluate the effect of various industrial activities on inland wetland habitat.

3.4.4. *Manual on-screen digitization of change*

A significant amount of high spatial resolution remote sensor data is available for change detection purposes (e.g., National Aerial Photography Program, *SPOT* 10 × 10 m). Some of these data are rectified and used as planimetric basemaps or orthophotomaps in geographic information systems (Jensen, 1995b). The aerial photography data often are scanned (digitized) at high resolutions, creating digital image files (Cowen et al., 1995). These digital photographic datasets are then registered to a common basemap and compared to identify change. Digitized, high-resolution aerial photography displayed on a CRT screen are easily interpreted using standard photo-interpretation techniques (size, shape, shadow, tone, color, texture, site, association). Therefore, it is becoming increasingly common for analysts to visually interpret both dates of aerial photography (or other type of remote sensor data) on the screen, annotate the important features using heads-up on-screen digitizing, and compare the various images to detect change (Westmoreland and Stow, 1992; Cheng et al., 1992; Ferguson, Wood, and Graham, 1993; Jensen et al., 1994a). The process is especially easy when (a) both digitized photographs (or images) are displayed on the CRT at the same time, side by side, and (b) they are topologically linked through object-oriented programming so that a polygon drawn around a feature on one photograph will have the same polygon drawn around the same object on the other photograph.

Even more high spatial resolution imagery is scheduled to become available during the next ten years if the proposed systems listed in Table 3.2 are actually launched (Stoney, 1995; Jensen, 1995a). These satellites will increase the amount of manual on-screen digitization of change that takes place.

An example of the use of on-screen digitization of change caused by Hurricane Hugo is demonstrated in Figures 3.5 and 3.6. Hurricane Hugo with its 135 mph winds and 20-ft storm surge struck the South Carolina coastline north of Sullivan's Island on September 22, 1989 (Boone, 1989). Vertical panchromatic aerial photography obtained on July 1, 1988 were scanned at 250 lines per inch resolution using a Zeiss drum microdensitometer, rectified to the S.C. State Plane Coordinate System, and resampled to 0.3×0.3 m pixels (Figure 3.5a). Aerial photography acquired on October 5, 1989 were digitized in a similar manner and registered to the 1988 digital database (Figure 3.5b). Image analysts then performed on-screen digitization to identify the following features (Figure 3.6):

• buildings with no damage,
• buildings partially damaged (black),
• buildings completely damaged (white),
• buildings that were moved (white arrows depict movement),

Figure 3.5. (a) Panchromatic aerial photography of Sullivan's Island obtained on July 1, 1988 prior to Hurricane Hugo. The data were rectified to State Plane Coordinates and resampled to 0.3 × 0.3 m spatial resolution. (b) Panchromatic aerial photography of Sullivan's Island obtained on October 5, 1989 after Hurricane Hugo. The data were rectified to State Plane Coordinates and resampled to 0.3 × 0.3 m spatial resolution. (*Source*: Jensen, 1996).

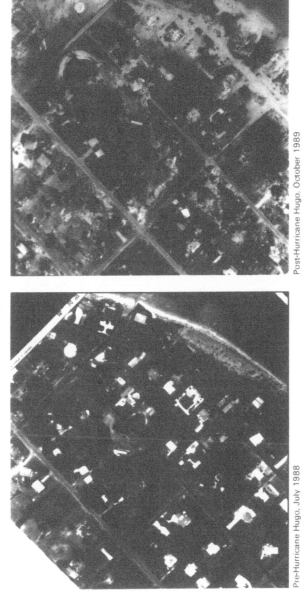

Sullivan's Island, South Carolina

Pre-Hurricane Hugo, July 1988

Post-Hurricane Hugo, October 1989

Scale

500   0   500   Meters

Figure 3.6. Change information overlaid on October 5, 1989 post–Hurricane
Hugo aerial photography. Completely destroyed houses are outlined in white.
Partially destroyed houses are outlined in black. A white arrow indicates the
direction of houses removed from their foundations. Three beach front
management set-back lines are shown in white (base, 20 year, 40 year). Areas
of beach erosion are depicted as black lines. Areas of beach accretion caused by
Hurricane Hugo are shown as dashed black lines. (*Source*: Jensen, 1996)

Sullivan's Island, S.C.
Change Detection After Hurricane Hugo

Scale

Meters

200   0   200   400   600

- buildings that might not be able to be rebuilt because they fell within certain S.C. Coastal Council beach front management set-back zones (base, 20 year, and 40 year lines shown in white),
- areas of beach erosion due to Hurricane Hugo, and
- areas of beach accretion due to Hurricane Hugo.

Several other types of change detection were applied to the digitized aerial photography, including post-classification comparison, image differencing, and write function memory insertion. Only the on-screen digitization procedure yielded accurate information on housing and geomorphological change caused by Hurricane Hugo.

3.4.5. *Image transformation (multiple-date principal components analysis)*

Numerous researchers have used principal component analysis (PCA) to detect change (Jensen, 1996). The method basically involves registering two (or more) dates of remotely sensed data to the same planimetric basemap as described earlier and then placing them in the same dataset. A PCA based on variance-covariance matrices or a standardized PCA based on analysis of correlation matrices is then performed (Fung and LeDrew, 1987; 1988; Eastman, 1992). This results in the computation of eigenvalues and factor loadings used to produce a new, uncorrelated PCA image dataset. Usually, several of the new bands of information are directly related to change. The difficulty arises when trying to interpret and label each component image. Nevertheless, the method is of value and is used frequently.

3.5. Digital change detection research directions

Performing accurate change detection using remotely sensed data is not a simple task. Below are a number of research areas that should serve to improve our ability to conduct accurate change detection:

- Scientists conducting digital change detection using remote sensing data and GIS data manipulation techniques must understand the *biophysical* characteristics of the soils/vegetation/water complex as well as the *cultural* land-tenure practices in the study area. Many change detection projects are doomed to failure because the scientists do not have sufficient physical and cultural science background.
- Additional work is required to accurately rectify multiple dates of off-nadir, remotely sensed data. Too often, scientists use rectification algorithms

designed to rectify nadir-looking, remotely sensed data on off-nadir imagery with only marginal results.

- Our atmospheric correction models must be improved so that the multiple-date remote sensor data are as radiometrically "clean" as possible for change detection. Works by Eckhardt et al. (1990), Hall et al. (1991), and Jensen et al. (1995a) that use pseudoinvariant radiometric ground control points to normalize multiple dates of imagery for change detection purposes represent a good beginning.

- A significant increase in the amount of hyperspectral remote sensor data is about to occur. Our current change detection algorithms can barely accommodate the multiple date analysis of just a few bands, e.g. <10, without serious problems. Improved change detection algorithms must be created that can process the multiple-date hyperspectral data.

- Jensen and Toll (1982) identified that change between dates may not always be compartmentalized into discrete classes. Rather, there may exist a continuum of change as a parcel of land changes from rangeland to fully developed, landscaped residential housing. Therefore, change detection algorithms of the future should probably incorporate some "fuzzy logic," which can take into account the imprecise nature of digital remote sensing change detection (Wang, 1990a, b; Jensen, 1996).

- New statistical theory is required to compute the accuracy of change detection products derived from the analysis of multiple-date classification maps. Initial work by Khorram et al. (1996) is promising.

Many GIS databases often contain dated landscape information. Up-to-date change information derived from remotely sensed data can be modeled with other information in the GIS to gain insight into the dynamic physical and cultural processes at work in the landscape.

4

Visualization and the integration of remote sensing and geographic information

N.L. Faust and J.L. Star

4.1. Introduction and definitions

Visualization is one of the most powerful technologies to evolve in the late 1980s and the early 1990s. Its power lies in the potential for direct integration into many disciplines, providing a tool that is seen not as a separate entity, but rather, one that is closely tied to analysis functions in each field.

Visualization is an overused and potentially ambiguous term that can mean almost anything having to do with the understanding of any type of data. It can mean a method of understanding the interrelationships between variables in a partial differential equation for thermal flow, but it can also mean an aid to show relationships between household income and crime over the last ten years in a metropolitan area. In this chapter we define visualization as techniques that aid in the interpretation of spatial data sets from remotely sensed data and geographic information systems. The common thread of GIS and remote sensing data is the geographic (x, y, z) location associated with all possible data values. For example, a point on the Earth's surface, indicated as a latitude and longitude tuple (x, y), might have multiple attributes such as elevation, landcover/landuse type, land ownership, census district, population per unit area, slope, etc. Many of these parameters may be indirectly related to specific measurements or based on the results of numerical models and simulations.

For traditional remote sensing data, an x, y, z position has multivariate data associated with the reflectance of sunlight in many discrete wavelength intervals; it might have a measure of the emitted radiation showing the thermal characteristic of an area or a measure of reflected radiation from a man-made source as in radar. As remote sensing systems get more complex, the amount of data associated with any area on the Earth's surface increases dramatically; this is easily seen in the proposed coordinated measurements of the Earth's surface and atmosphere, at various wavelengths, over a two-decade time period from the Earth Observing System (Lu, 1992). Sensor-derived data are normally dependent on season (Miller et al., 1991), time of day (Lillesand and Kiefer, 1987), the geometry of the observation (Kimes, Smith, and Ranson, 1980), and weather conditions, so variability of remote sensing data compounds the visualization needs for spatial data.

Data measurement capabilities provide a wealth of data on not only the Earth's surface but on the atmosphere above us and the Earth volume below us. As we strive to understand the causes and effects of change in our environment, we must better understand measured data, develop models of environmental processes, and iteratively improve those models as we gain knowledge. With the current emphasis on the environment we have a dramatic need to improve the tools with which scientists, technicians, politicians, and the general public interpret the massive data volumes collected by advanced sensor and computing technology.

Potentially, visualization of spatial phenomena is a natural extension of experiences in a person's normal life, and it should be easier to understand the relationships between variables when they are presented within a geographic context than without. For example, the NO_2 pollution concentrations over an urban area might be easier to comprehend when superimposed on a map of the city (Figure 4.1, see www.cambridge.org/9780521158800). The human mind works well on spatial correlations, and a coincidence of high NO_2 levels with the sites in a major industrial complex in the city are unlikely to be overlooked. Viewing NO_2 concentrations over the eastern United States in a time series clearly shows that emissions often are channeled by weather patterns up the eastern coast along the Appalachian Mountains. Viewing this time sequence in three dimensions further demonstrates the pollution transport (Figure 4.2).

This chapter addresses trends in visualizing spatial data. Even though there are ample examples in medicine, we emphasize the visualization of atmospheric, subsurface, and landscape features at various scales. Our overriding view is that visualization tools are an important component in the integration of remote sensing and geographic information systems. In addition to the database, applications, and modeling viewpoints represented in other chapters in this volume, visualization can be an effective way to increase the bandwidth between the computer systems, which store and process spatial data of many kinds, and the individuals who wish to understand and work with these data.

4.2. Types and techniques

4.2.1. Two-dimensional data

One of the simplest forms of representation of spatial data is a gray-scale image of a single variable, which contains information about every image picture element (pixel) as well as the inherent spatial behavior of the variable. Without losing generality, each element in a vector data structure (point, line, and area) may be treated the same way; the information about the vector element is presented in gray-scale. In most images a considerable part of the value of this presentation is not only knowing the data value (as indicated by the gray-scale value) at a location, but also in knowing the spatial relationships between one location and others in its neighborhood. The eye has extraordinary abilities for recognizing spatial relationships. The fact that a location in an image has a high value (perhaps indicating high reflectance at 500–600 nm) might lead one to believe that the pixel consists of concrete. Identifying that pixel as part of a road depends on relating that pixel to its neighbors and the recognition of a linear pattern in the image. Similarly, for traditional GIS data types, an area of high values may represent high migration rates over some time interval;

Figure 4.2. Pollution density as a three-dimensional, time-dependent process

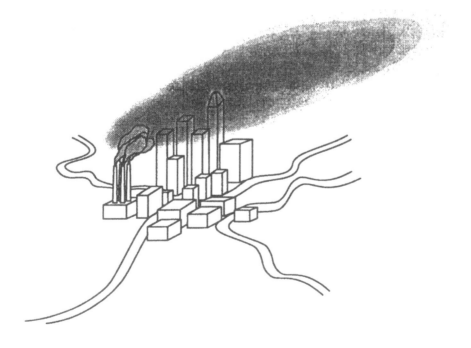

the spatial pattern of the variable might permit one to examine the hypothesis of net migration out of the countryside into an expanding urban core.

In remote sensing systems, data typically may be quantized into 16 levels (4 bits), 256 levels (8 bits), 65,560 levels (16 bits), or greater. For example, the standard products from the *Landsat* and *SPOT* satellite programs are 8 bits per pixel. Similarly, traditional GIS datasets vary in number of discernible levels and storage. Elevation, whether stored in raster or vector formats, may require 16 or more bits in areas of wide elevation range. For example, the digital elevation dataset covering the Santa Barbara quadrangle in California has a minimum of 0 and a maximum of 1,209 meters above datum, requiring at least 11 bits of storage per element for 1 meter of precision. Demographic data is typically stored as categorical variables; as an example, rates of population increase in one atlas (*Times Atlas of the World*, 1981) are presented as one of four categories of annual change, and population density as one of nine categories.

No matter what the dynamic range of the data may be, the discrimination potential of the human eye sets a limit in the interpretability of a gray-scale presentation. The human eye cannot distinguish more that 64 gray levels in an image. The ability of a human to accurately interpret and find features in a gray-scale image may be a function of local contrast variations and image structure (Adelson, 1993). Many computer displays, including notebooks and laptops, can display 64 to 256 gray shades (Faust, 1991). If the input data range is less than that of the computer display system, an image enhancement technique called contrast stretching may be used to map the original data contents onto the gray-scale capability of the observer, and in this way, effectively present the dynamic range of the data to the user. More on various contrast stretching enhancement techniques may be found in the *Encyclopedia of Computing* (Faust, 1989) or in image processing texts (Showengert, 1983, Lillisand and Kiefer, 1987). Hardcopy media will not be discussed in this article. For more information, see the *Manual of Photogrammetry* (Slama, 1980).

Single variable data and images may also be mapped into colors. The data values are mapped through a look-up table into a specific color set that can often be tailored by the user; this permits the display of more categories of information than is possible with gray-scale methods and is called *pseudocolor mapping*. One should be careful because the human eye responds differently to some hues than it does to others. For example, reds and yellows tend to grab the attention more than blues and violets. In addition, the relationship between gray scales in a gray-scale image is linear and obvious, whereas in a pseudocolor image the relationships between the human response to assigned colors is complex.

The representation of multiple variables in a single portrayal is extremely common in remote sensing applications. For example, in the case of *Landsat* TM data, there are three bands that record data in the red, green, and blue regions of the spectrum. A direct mapping of these three bands of TM data to the red, green, and blue color guns on a normal color monitor should give a true color image very similar to what a person would see if he or she were sitting in the satellite and looking at the Earth through a telescope.

In fact, however, the *Landsat* sensors, and those from other nation's satellites and aircraft, create images representing the Earth in many regions outside the range of human vision. For example, the TM sensor collects data in seven bands, with only three of the bands corresponding to human vision. These bands contain data that can be related to vegetation health, mineral composition, and thermal emissivity of the Earth's surface (Colwell, 1983).

Part of the visualization problem in remote sensing is to present these multivariate spectral data so that one can correctly interpret spatial relationships between the components of the Earth's surface. This is a difficult problem in that the analyst must learn certain characteristic traits that are represented in the spectral regions outside of human visual range and try to apply them to color images that are within visual range – a fundamental component of visualization. In remote sensing practice, there are a number of well-understood transformations that attempt to extract features about the Earth's surface through transformations of the input spectral bands (Kauth and Thomas 1976; McFarland, Miller, and Neale, 1990). As more and more spectral bands become available on multispectral and hyperspectral sensor systems, interpretation becomes potentially more complex.

This multivariate data fusion is less common in GIS applications. One could imagine a set of data on population through time as a function of location wherein the red, green, and blue color components could correspond to three successive census epochs of the population. In this case, any location appearing in a shade of gray would indicate no change in population through the time period, red tones would indicate population declining over the interval, and blue tones would indicate populations increasing over the interval.

It is often hard for a person to understand colors in a red, green, blue (RGB) color space (Haydin, Dalke, and Henkel, 1982). However, a person does relate well to a color space in which the color axes are intensity or brightness, hue, and saturation (IHS). There is a direct relationship between IHS and RGB color spaces, and most visualization systems permit these conversions.

The next technique for visualization of two-dimensional (2D) data sets actually uses information from a second registered dataset to enhance understanding. If a dataset and a digital elevation model are geometrically registered, an elevation data point may be derived for each image display pixel. With the Sun at a specific elevation and azimuth, terrain shadows may be cast using a dot product of the Sun vector to the normal of each facet of the elevation dataset. A facet consists of three adjacent elevation data values, which define a plane that connects the vertices. Knowing this geometry, a normal vector may be calculated for terrain facet. Each terrain facet is compared to the Sun direction vector; those facets that are facing away from the Sun are darkened to represent shadow, while those facing the Sun are left with the same color values. This technique of relief shading gives raster GIS data a pseudo-three-dimensional appearance since information is present about the terrain elevation as well as the GIS image. Figure 4.3 (see www.cambridge.org/9780521158800) shows a raster GIS image with no shading, contrasted with the same image after relief shading has been applied.

Another way to increase the information content of a portrayal of spatial data is to overlay two or more themes. Most thematic maps take this approach, combining (for example) elevation, vegetation, transportation, and political boundaries. Commonly, we enhance the interpretability of imagery datasets by adding GIS information as an overlay on top of the image. Although imagery in general represents continuous gray-scale or color data, most GIS data are often categorized. By overlaying GIS variables, relationships between image data and discrete data provided by the GIS can be developed to assist users in their task (Figure 4.4, see www.cambridge.org/9780521158800).

A common approach when multiple sets of two-dimensional data represent instances through time is to animate the graphic presentation. The multidate population data example above is a natural in this regard: Since our eyes are very sensitive to change, these animated sequences are very effective for the portrayal of change.

In this regard, one of the principal uses for remote sensing data is the study of vegetation and cultural features at different points in time – different times of the day, different times within a single year, different times across years (e.g., Figure 4.5). As in the animation of more traditional GIS datasets, the images must be geometrically corrected to the same coordinate system and scale; there may also be a need to correct for view and illumination geometry as well. In some cases images will be classified to create a GIS layer showing land cover classes; in other cases continuous variables such as a vegetation index may be calculated. Change detection using classified imagery also may

Plate 1. (Figure 4.1.) Superimposed pollution density and urban features

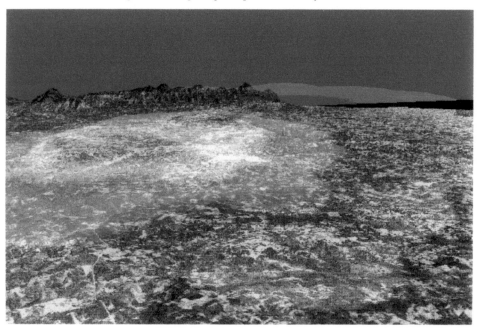

Plate 2. (Figure 4.3.) Raster GIS image, with shading based on a coregistered elevation dataset

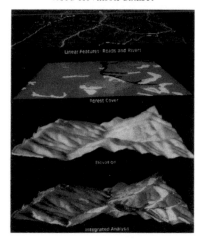

Plate 3. (Figure 4.4.) Merging image and thematic GIS datasets for South Georgia flood study

Plate 4. (Figure 4.6.) Combination of two geophysical datasets using 2.5 dimensions

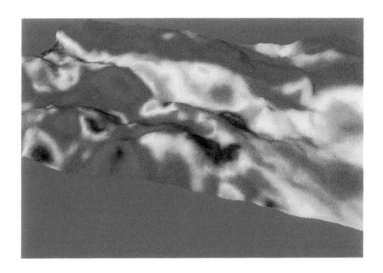

Plate 5. (Figure 4.13.) Virtual interaction with GIS database using 3D rendering

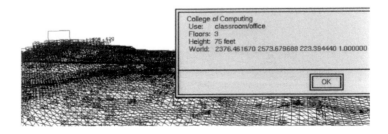

contain information for each pixel as to the original class and the new class. In complete generality, the temporal sequence may be made up of merged 2D views of imagery overlain with GIS datasets (such as TM data with superposed political boundaries and symbology for transportation networks).

4.2.2. 2.5 Dimensions – surface rendering

One of the most natural ways for a user to interface with image data and GIS information is to show the datasets as if one were looking out a window from a selected viewing site. If the data are provided in this context, a user can apply a lifetime of experience to aid in the interpretation, in contrast to the

Figure 4.5. Representation of an animation to illustrate population growth through time

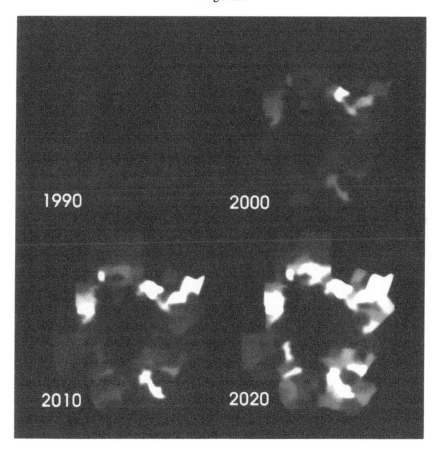

artificially limited perspective of 2D techniques discussed in Section 4.2.1. Rendering algorithms have been developed that can portray a database and an elevation dataset together as a perspective image. This process is known as two-and-one-half-dimensional (2.5) rendering. The process is not considered full three-dimensional rendering because the elevation and image databases are inherently single-valued functions of the elevation z. For example, a beach ball will have one elevation value for its top and another for its bottom, and thus it requires a 3D treatment. A 2.5D rendering is of significant value for visualizing GIS layers in relationship to elevation data. It may also be used to relate other sources of data where one data source is taken as the color variable and another is taken to represent elevation data. Figure 4.6 (see www.cambridge.org/9780521158800) shows a 2.5D rendering of a combination of two geophysical datasets. In this case, gravity data is treated as the topography and color data is taken from magnetics data. Unfortunately, in many systems the rendering process for a single graphic may take thirty minutes or more, depending on the size of the database. Many vendors have over-simplified the database to achieve real-time capabilities for rendering. If a compromise is forced, the rendering will look artificial and will not achieve the necessary realism. GTRI (Georgia Tech Research Institute) has recently developed a very efficient rendering code, optimized with multiple resolution aliasing logic and shading techniques (Faust, 1989). This code is capable of producing fully realistic rendered fly throughframes at a speed of 10 seconds per frame (on a Sun SparcStation 10), thus allowing for the rapid generation of complex pre-planned flight paths as well as the interactive preview of wire frame and interpolated shading scenes.

The procedure used in the rendering goes from a source database to the screen, not from the screen to the database. Generally, the elevation and the thematic database are processed to have the same size image pixels. The databases are read a line at a time and triangles are constructed between adjacent lines. The order in which the database is read and drawn is top to bottom, left to right – no matter what the viewpoint and field of view. A volume is projected on the database showing which parts of the database a viewer can see. Using the limits of the view volume, optimized access to the database is attempted. For each facet of the terrain, the corners of the facet are passed through a perspective transformation to get screen coordinates for the projected facet. Shading is performed relative to the thematic values at the three vertices of the triangle.

The 2.5D rendering uses a technique called Z buffering to determine which parts of a database fall in front of other parts of the database and are thus visible. This approach allows objects that are nearer to the eye to occlude objects farther away. The rendering techniques described above use no external light sources.

Ray tracing is another approach to rendering and has been used to create high-quality rendered images, but often at a huge computational cost. Ray tracing runs have been known to take days on a supercomputer for a single complex scene. Individual rays are cast from each light source to the object database. Given the direction and distance along the source ray, an intersection is found between that ray and the terrain database. At that point, the ray is reflected in a direction dependent on the incoming ray angle and the normal to the terrain facet. In the simplistic case of only one reflection, the reflected ray is tested to see if it intersects with the viewplane of the user. If it does not intersect, the ray is ignored. A large number of rays are cast to eventually build up a rendered image.

To be able to calculate this reflection, every facet of the database and every object must possess characteristics that enable one to estimate the specular versus diffuse aspects of the surface, as well as to perform mappings between thematic contents and color. In a GIS database, one might assign color and reflectance characteristics to specific objects or categories in a sense in the same way that a cartographer makes choices for display on a map. Remote sensing images do not have this information available; thus a pattern recognition technique must be applied to determine the classification of every terrain facet. Then, the physical understanding of the material must be available to estimate the amount of transmission, reflection, and refraction that occurs based on incidence angle.

Successful ray tracing examples usually involve detailed models for both every object and every material within the scene. Multiple reflections are used where the ray may not directly intersect the viewplane, but may be reflected onto another object or terrain facet. This secondary reflection is then traced to see if it intersects another object or comes to the view screen. Beautiful examples of ray tracing have been created where reflections of objects or terrain show in a window within the observer's field of view (Siggraph, 1989).

Two algorithms have been developed to decrease the calculations required to estimate ray tracing solutions. Inverse ray tracing follows a ray through a 3D scene, tracing a ray from the observer's screen and calculating where it intersects the terrain. By keeping track of where the last few rays intersected the terrain, a small search radius is used in calculating the intersections rather than testing all objects in the scene. *Radiosity* is a rendering approach that precomputes the reflectance for a terrain or object facet for all possible combinations of angular intersections of source light ray with the terrain

facet. This technique requires a modeling of terrain and objects and a sufficient knowledge of light/terrain interaction to be able to compute the radiosity
view matrix. It has been successfully used in the rendering of complex scenes.

Several types of shading can be accomplished with most of the rendering techniques described in this section (Finlay, 1993). Constant shading
generally involves taking the color values at each facet vertex (three for triangular facets), averaging the color, and placing the average at each pixel
in the screen within the projected triangle facet. This technique is fast but
often leaves objectionable artifacts between facet edges close to the viewer.
In Gouraud shading one interpolates colors between the corners of the projected facet based on the distance from each vertex. This generally provides
smooth shading with no artifacts or discontinuities at the edges of facets and is
the most often used technique in high-quality rendering of terrain databases.
Phong shading requires that a normal vector be calculated for each terrain
facet. The angle between the source and the normal is then obtained from the
dot product of the normal vector with a source vector. Based on this angle, a
reflectance is calculated that is based on the source intensity and bidirectional
reflectance characteristics of the terrain or object material. Phong shading is
the most time consuming of the shading techniques listed, but potentially the
most realistic.

4.2.3. Three dimensions – volume

Full rendering of three-dimensional scenes can be a complex and time-
consuming process. Two approaches will be discussed that provide adequate
renditions of 3D databases and themes and objects.

One of the main problems with rendering remote sensing and GIS data
sets is that the underlying data structures are inherently two dimensional.
Each pixel in a GIS will have an attribute assigned to it in a gridded system,
and each point, line, and polygon will have an attribute assigned to it in a
vector/polygon-based GIS. Three-dimensional data structures designed for
efficient manipulation are an active research area. Recent developments in
object-oriented databases may have great impact on this application, although
these are not commonly available (Stonebraker, 1990).

Voxel rendering is an extension of techniques used for surface rendering. A
voxel database is a 3D grid with constant increments between x and y positions
in a 3D coordinate system and constant increments in the z direction. As
opposed to the surface rendering algorithms where different subsurface layers
may be represented as several 2.5D databases, the voxel rendering technique

requires that a database value exist for all discrete values of x, y, and z in the volume. In most cases it is very difficult to obtain measured data throughout the total volume. Models, however, can be used to generate data at every voxel given appropriate boundary conditions. For example, in a representation of a subsurface water body, attribute values such as modeled flow velocity and direction are available for every voxel. When measured data are available, they are usually provided only at certain discrete points within the volume. To create a voxel database, these data are interpolated (Finlay, 1993).

A number of techniques are in use for rendering voxel databases. For a data volume, one of the easiest methods to visualize data is a slice in any dimension. One of the axes will be fixed and an image will be shown of that slice through the database (Figure 4.7). If the entire database is in computer memory, this rendering may happen very fast. In enhancing the slice, functions such as gray-scale and pseudocolor display are common.

Another technique is to define arbitrary start and end points for transects through the data. For each point on a line connecting the end points the database is queried and a vector of data is acquired with one value for each z in the voxel database. The resulting product is a graph where the horizontal direction represents distance along the transect, the vertical dimension represents the z direction, and the color represents the variable value at that point. This type of display is useful to emphasize differences in data values when crossing boundaries.

A third technique allows visualization of two transects perpendicular to each other. The view would normally be looking directly at the edge of

Figure 4.7. Slice through a three-dimensional dataset

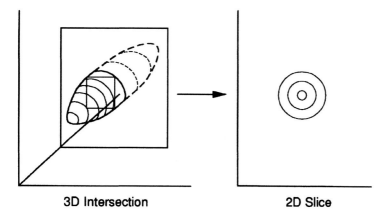

3D Intersection   2D Slice

the intersecting transect planes. If the data volume is thought of as a cube, we would be looking at an edge of the cube (Figure 4.8). View points and intersection points could be manipulated by the user. For example, another

Figure 4.8. Presentation of three-dimensional data as a
cube (courtesy NASA, 1987)

view might be from an elevation angle of 45 degrees and an azimuth of 130 degrees.

The most general technique is full rendering of the volume in perspective. A viewer may be placed either outside or inside the volume and may look in any direction. Techniques of translucency may be used to allow the user to see through voxels with selected data values while having other voxels be opaque (Foley and Van Dam, 1984).

4.2.4. Four dimensions using time

One of the more innovative visualization areas is the analysis and display of three-dimensional data sets that change over time, as an instance of a four-dimensional data visualization problem. The data normally would consist of x, y, and z data in a voxel data structure, numerous layers of 2.5D data combined with 2D overlays, or part of an object-oriented data structure. Two types of presentations of these data are possible. Eulerian visualization fixes a coordinate at a viewer's location and allows the movement of data within the observer's view volume. In Lagrangian visualization, it is the viewer that moves through the data. The coordinate system of the viewer must be related dynamically to that of the spatial data base. In either case, the data are presented in a perspective or orthogonal projection.

The simplest form of 4D visualization might be a view of a voxel database while slices for varying x coordinates are taken dynamically. The slices could be along one of the spatial axes, or they could be taken along any direction through the database. This type of presentation is often used to enhance the display of underground features for hydrologic and geologic investigations. Instead of the viewer moving through the database, as in Lagrangian motion, the observer stays fixed, and the database is peeled back in small finite steps.

If data are available from models or from measurements, a fixed viewer could observe a data volume using voxel or surface rendering tools. The volume could be rendered with the observer only seeing the outside edge or an internal slice into the data cube, or parts of the volume could be rendered translucent so that an understanding of important regions within the data set is enhanced. In the presentation of time sequences, the relationship between the displayed sequence of images and the time frame at which the observations or model steps were taken must be explicit. At the present time dynamic modeling of each time step is only practical on supercomputers; common practice is to record the image components digitally and retrieve them later. Well-known applications of this technique include the real-time generation or playback of fly throughs of terrain data (real or synthetic). These functions are inherently Lagrangian in form since they include an observer traveling through a geographic space. Other objects within the space also have the ability to move. Flight simulation companies have been developing such databases for training, often based on specialized hardware; Evans and Sutherland and Singer/Link were major players in these developments.

In addition to the Z buffer innovation already mentioned, techniques called *texturing* and *phototexturing* were developed. These allow the mapping of a modeled mathematical texture or actual digital photographs to terrain or object polygons. Using these techniques, vastly improved rendering capabilities were developed that provide very realistic representations of the terrain and objects, while still only requiring the rendering of a finite number of vertices (Figure 4.9).

Current simulations being generated by the Environmental Protection Agency in Research Triangle, North Carolina involve traveling through a geographic database and a combination of terrain visualization and the display of data representing air pollutant concentrations (Rhyne, 1992). In another version, a perspective image of the terrain is displayed and a time sequence of atmospheric pollutant transport is overlaid as a translucent layer with colors representing the relative concentrations. A voxel method showing model or

Figure 4.9. Rendering with texturing and photo-texturing

measurement data may be combined with 2.5D rendering of the terrain with concentrations shown in varying shades and degrees of translucency. An x, y, or z slice could also be used to view discrete parts of the datasets (USGS, 1992).

One of the most interesting techniques used by these researchers employs sound to notify viewers of an approaching rainstorm. Not only can the rain front be seen approaching by changes in pollutant concentrations due to the capture of particles by the rainfall, but sound allows an observer to investigate the overall dataset while keeping track of the frontal system. This application of multimedia technology combined with rendering is effective in the analysis of complex natural phenomenology. Although the models in this case involved large-scale computing on supercomputers and was not performed in real time, the results of an action sequence was captured in video and used in a playback mode.

4.2.5. N dimensions

Sophisticated higher dimensional visualizations are relatively uncommon in spatial data analysis applications. Although IGIS research commonly includes numerous variables, common practise is to use thematic maps to portray the variables. For imaging sensors being developed that have potentially hundreds of spectral bands (NASA, 1987), as well as for the analysis and interpretation of complex multivariate GIS databases, a method must be devised to allow the user to interpret a massive amount of information unlike any previously experienced source.

The analysis of hyper-dimensional data may be assisted by visualization techniques that combine 2.5D surface and 3D voxel rendering with time sequences in which the variables are overlaid on the terrain in various combinations that may change by type of feature being investigated. As one approaches a wilderness area, the user could freeze the view and index through previously selected spectral band combinations designed to enhance vegetation differences; over densely populated areas combinations of color, texture, or statistical graphics would portray differences in population age structure. Modern visualization hardware could probably support such a synthetic presentation, although current software could not.

4.3. Applications

Because visualization is simply the process of making data more interpretable, it has application in all of the disciplines normally associated with remote sensing and GIS analysis. The image display and enhancement functions normally considered part of remote sensing are actually visualization techniques. The availability of multispectral datasets around the world at lower costs than in the past provides opportunities for innovation in understanding the global environment. However, the availability of this massive amount of data will tax even the most capable of the current commercial remote sensing and GIS systems. Visualization is critical in assembling and analyzing these data. The tools must be made available to the applications specialists so that they can find the capabilities intuitive in nature, and these tools must be easy to apply to their problems.

Although visualization is currently the buzzword in the research community there is a need to provide these tools to scientists, applications engineers, enforcement specialists, educators, and political officials so that advances in data availability, understanding of global models, and assessment of potential environmental effects can be exploited in protecting the only Earth that we have. We are at a crossroads where technology may have a direct and positive impact on our future generations if we take advantage of it. In this section, we briefly review a few applications areas that are progressing rapidly and that involve spatial data and spatial data processing.

4.3.1. Landscape

Landscape analysis and planning was one of the first disciplines to develop and use geographic databases. Landscape architects at Harvard motivated the development of what might be considered the first grid-based geographic information system software that was not directly controlled by a complex model (Sinton and Steinitz, 1971). Landscape architects generally fall into two broad categories that differ principally in the scale at which they work. One group develops detailed designs for individual lots for single and multifamily housing. They are intimately involved in the selection of plants, bushes, and trees that would enhance the appeal of the yard. This group also extends to design of environments for buildings, shopping centers, etc. The other group focuses more on regional planning, evaluation of large-area terrain, and the combination of spatially related data sets. This group is extensively involved in research and application of the current and next generation GIS systems.

One of the key factors to landscape architects is the visual appearance or quality of a landscape (Figure 4.10). Landscape architects are direct users of the visualization techniques discussed in Section 4.2. Innovative techniques discussed in Section 4.6 will provide a tool for landscape architects that will allow complete evaluation of measures of visual quality.

4.3.2. Flight simulation

Flight simulation developers have always used leading-edge technologies for the visualization of aircraft flying above a natural terrain. Developers have special purpose hardware to render in real time. Realism is extremely important in this application. High-end systems normally cost 3 to 8 million dollars and have extensive computing power, high-quality graphics, motion simulation systems, and sound-based cues. These developers were the first to implement phototexture to add dramatic realism to their simulations. In addition, concepts of parallel rendering were first theorized, designed, and implemented on flight simulators.

Flight simulation companies are in an extremely difficult situation now that some of the new hardware for workstations has been introduced. The workstation implementation of phototexturing and parallel processing brings these systems close to the real-time world of the big simulation manufacturers. These machines cost at least a factor of ten less than those of only a couple of years ago, but have almost the same capability.

4.3.3. Architecture

For many years architects slaved over diagrams, floor plans, and final renderings using drawing paper, straight edges, and T squares. The architecture and engineering professions have motivated major development of computer aided design (CAD) tools. CAD models of buildings allow the interaction

Figure 4.10. Rendering for landscape architecture

beween a potential customer and the architect in a fast and efficient manner
(Figure 4.11). However, there is some concern in the architecture community
that the automated design tools in some cases make it harder to envision new
concepts than the traditional drafting board. For an architect, there is an
essential element of art in design that is not well nurtured in a CAD environment. The trend toward more and more capable simulation of a real world
environment using workstations and phototexture will probably have a great
influence on the use of the new tools by the architecture profession.

Figure 4.11. Visualization based on CAD models of buildings

4.3.4. Sensor fusion

The ability to take signals from a diverse set of sensors, either colocated or separate, and to combine the information together into a single data stream that can more easily be interpreted by a human observer or a computer algorithm is a capability that will tax even the most robust of visualization systems. Many of these sensors are fundamentally different from others in a multisensor suite. For example, multispectral imagery responds to chemical or biological parameters, whereas radar responds primarily to physical parameters such as surface roughness and dielectric constants. Thermal radiation depends on emissivity. The information content of these three types of imagery is distinctly different, and it is difficult to show correlation between the images even though they may be of the same terrain. Visible sensors such as vidicons or color cameras take imagery in a manner known as angle/angle sensing. The sensor, perhaps a CCD detector, records radiation from equal angular regions even though the area covered by one image pixel varies throughout the field of view. Infrared sensors have the same angle/angle representation. A radar sensor, in contrast, records information in a range/crossrange manner where every image pixel has the same area of coverage. It is difficult to determine the correlation between what is seen in the visible and infrared bands and what is seen in radar. New visualization techniques are necessary to allow better understanding of the integration of multiple sources of data as well as to take these remote sensing systems and merge their information content with traditionally GIS data such as corridors of mobility and visibility.

4.3.5. Aerospace

Aerospace engineers have been using visualization for a long time for complex analyses. More than just a CAD model, an aircraft being designed has to be analyzed for structural integrity by looking at stresses applied to the model. Finite difference techniques are used to analyze a model's reaction to linear and some nonlinear stresses. Visualization has been used with finite difference implementations of stress partial differential equations. Partial differential equations for equations of motion are implemented in finite difference equations and model reaction can be viewed as a function of speed. Aerospace engineers have been using wind tunnel tests for years to gauge the potential advantages that might be gained by using a certain design. Visualization techniques may be used to compare measured wind tunnel data against the finite difference approximation; one might argue that this comparison is fundamentally one of

3D GIS technology, fusing all possible input datasets based on spatial location and comparing this potentially tensor volume against models.

4.3.6. Atmospheric data

Major efforts are now being focused on the state of the environment of the Earth. Massive amounts of atmospheric data are being gathered and models are being developed to try to understand the parameters that affect the system. Atmospheric data are inherently 4D data: x, y, z, and time. Visualization techniques are being used to show the time and space dependence of the atmosphere; however, new techniques will be needed to manage the volumes of data from measurements and models and to understand the differences between the real world and our simplified models of the Earth system.

4.4. Hardware considerations

Hardware capabilities of various computing systems have much to do with the potential use of visualization for a particular project. If a system has only minimal computer graphics capability, such as a gray-scale display, the use of visualization may not be warranted since it would have to be shown in a much more restricted environment than the visualization technique demands. For effective use, visualization techniques must be able to adequately represent the measured data or model results. A 16-color VGA display on a personal computer may not be enough.

In addition to the graphic part of the system that provides the direct interface with the user, the available computing power is a significant issue for visualization. Visualization is not common in many disciplines because of the amount of time necessary to graphically represent the desired data. Several hours for creation of a perspective view is simply unacceptable to both research and applications personnel. New advances in computing power have placed tools previously available to only a small, sophisticated, set of researchers in the hands of a wider circle of application engineers, teachers, and management teams. Dynamically interacting with your dataset by providing different types of graphical presentation in real time is now demanded by users of spatial data. Space does not permit a detailed review of hardware considerations; we touch on a few essential points.

The capabilities of computing systems have been dramatically changing over the past few years as new chip technology has caused major improvements in the processing power that can be cost effectively applied to real-world problems. Since 1986 the power of computing has doubled every year. A model for this speed improvement is given by Joy's Law (Gage, 1990):

$$\text{speed (in MIPS)} = 2(\text{Year} - 1986),$$

where MIPS is millions of instructions per second that can be processed by the computer system. Figure 4.12 shows that we have followed Joy's Law very closely, but are now exceeding the rate of speed increase. In 1986, the Digital Equipment Corporation VAX 11/780 was developed; it had a 1 MIP rating. Current workstation central processing units are now reaching between 100 and 150 MIPS. For visualization, another factor is the number of floating point operations per second that can be accomplished since many floating point operations are normally required. The number of millions of floating point operations per second for a system is normally given in Megaflops. Hewlett Packard has recently announced a 150 Megaflop processor that will drive the computer industry toward greater floating point capability as the competition for engineering and visualization workstations increases. Silicon Graphics has also announced a 300 Megaflop system for their Onyx workstation series. This is supplemented by their "Reality Engine" graphics.

Figure 4.12. Plot of Joy's Law (after Advanced Imaging, Jan. 1993)

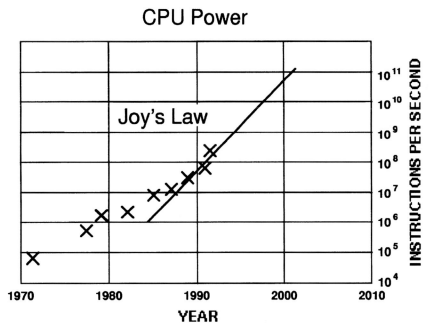

Advanced computing power lies not only in the speed of a particular CPU chip, but even more so in the implementation of parallel computing in workstations. Parallel computing allows one to break up serial processes within programs and assign processors to specific parts of an algorithm. Language compilers and operating systems should provide this capability of automatically creating threads for parallel operation by analysis of the code.

It is hard to look in a crystal ball and tell the future of the computing industries' venture into high-capability, low-cost workstations. However, it appears that we will continue to advance in computing speed at a rate equal to or greater than Joy's Law. Inexpensive parallel computing with shared memory will turn visualization from a scientist's plaything to a workhorse for the rest of the world.

4.5. Software systems

Software for visualization is provided on commodity personal computers as well as most UNIX workstations. This visualization capability is generally the simplest type in which an image or computer graphics can be shown, edited, and saved in two dimensions. Some three-dimensional software is available on some of these systems. Four-dimensional visualization is rarely seen but is currently being used in research. The challenge in software is to fully integrate visualization tools into the application environment. One should not have to leave an image processing or GIS environment in order to "visualize" what the raw data or model is saying. The linkage between the applications and the newly developed tools is tenuous at best and needs to be strengthened both from the application side and from the visualization techniques side. A review of software systems that support visualization is outside the scope of this chapter and would be out of date by the time you read this. Conferences such as SIGGraph (and its proceedings) are excellent ways to survey the state of the art in visualization software, although they rarely focus on the issues of integrating GIS and remote sensing.

In the research area, the Georgia Institute of Technology (Georgia Tech) is developing a highly efficient system for the rendering of very large 2.5D scenes in support of this integration (Faust, 1989). This work to date has not been developed under contract to the government, so rights for the software reside with the university. Improvements to the model are currently underway that will allow experimenting with real-time rendering of geographic databases. These techniques are likely to lead to a whole new perception as to the appropriate interface for spatially located image and GIS data.

4.6. Future directions

Because of major advances in hardware and software technology in the past several years, the concepts of how remote sensing and GIS data are dynamically linked is changing. Some of the new concepts now being implemented could not have been successfully pursued even as recently as five years ago. In many ways this is similar to the dilemma of finding the cost/benefits of any new technology. Part of the equation is finding out how you can do the things that you normally do in a faster and more cost effective manner. However, the principal advantages of the technology advances lie not merely in achieving faster results, but in being able to perform new functions that could not have attempted before.

Visualization meets both of these criteria. Not only does it allow a quicker and cleaner presentation of standard analyses, but it allows one to define new analyses and functions that could not have been envisioned before. Current visualization efforts have been used to demonstrate analysis results to fellow scientists, laypersons, and members of the political infrastructure who need those results to determine future policy actions. This is an important and undervalued use of the visualization tools that are being developed. Far greater value will accrue in using these new techniques to find new ways to solve previously unsolvable problems. In the purely scientific domain, effort is being focused on the application of parallel and supercomputer technology toward the solving of "Grand Challenge" problems. In the applications domain even greater advantages may be gained due to these advances that apply to the solution of real world problems.

One of the rapidly expanding tools for visualization that has found numerous practical applications is geographic information systems. GIS is a multi billion dollar industry that is poised for exponential growth. No longer are GIS systems only in universities and research institutions, but now the GIS technology is being used at all levels of government for the management of critical spatial datasets. By having such a diverse user base, GIS has forced commercial vendors to produce easy to use and intuitive software for the management of geographic data. This easy to use implementation, of course, is not complete since requirements change faster than the developments that commercial, university, and government entities can be completed. As mentioned earlier, we believe that the future lies in IGIS: an integrated GIS where remotely sensed data may be a full partner in database creation, update, and analysis.

A concept for a "virtual 3D GIS" has been developed that takes advantage of the strengths of current GIS systems and tools and integrates visualization techniques usually thought of as being used in computer graphics into the dynamic analysis of spatial data. Because of the advances in computer power with fast RISC chips, ASICs, and parallel computing, and also due to the development of highly efficient techniques for 2.5 and 3D perspective rendering, users may totally immerse themselves within a virtual environment built around 2D and 3D GIS data. By placing the user within the spatial database and performing functions that humans would normally do, a user may dynamically approach the solving of everyday problems, much in the same manner that one would normally react with the surrounding world. By performing standard GIS analyses totally in the virtual domain, a user can interface with the world in a very natural manner (Figure 4.13, see www.cambridge.org/9780521158800).

For example, if one is performing a GIS analysis to determine where to optimally sight a vacation home, one would not only want to know information on surrounding land use, population, land values, etc., but would also need to know what the view would be from a second story window on the south side overlooking a lake. The combination of normal GIS functions with a visual perception based interactive interface will allow a much more satisfying analysis to be made that will have conclusions that one can physically relate to. In addition, the portrayal of the GIS analysis as it is being performed in a totally 3D environment will allow for instantaneous feedback from siting studies, which might involve the proximity to water, location of steep slopes, and the coverage of the 100-year flood plain. As the areas are delineated, the 3D perspective shows the high slopes on hillsides, the location of streams and rivers at the bottom of the terrain valley, and the flood plain overlay on the flat areas next to the streams. Now a total analysis can be accepted more easily because the user is an integral part of the analysis and can immediately see and understand the implications of the GIS analysis. The user can instantaneously query directly in the perspective image and find out such information as "who owns the land on that neighboring hilltop" or how would building a dam on my property affect those downstream from my land (Figure 4.14).

With detailed information that might be derived from CAD modeling and GIS, a user would be able to ask what a visitor could see from the top of the Washington monument, or, in the case of Atlanta, what could Olympic visitors for the 1996 Olympic Games be able to see if they climbed to the top of Stone Mountain. The interactive query of geographic coordinates in perspective images as well as the retrieval of multiple descriptive attributes is a powerful tool that may very well be the interface of the future for integrated remote sensing and GIS systems. The 3D view would also be instructive in understanding hydrologic models or in estimating the shortest route using traditional network analysis. The user could see the network being built dynamically from a starting point to the destination, see the effects of possible traffic blockages, notice the effects of narrow roads along the foothills in channeling traffic in a certain manner, and figure out how a view from a window might have permitted the consideration of alternative routes.

Georgia Tech has been funded in a collaborative effort to develop a real-time 3D virtual reality GIS prototype for military use. The Georgia Tech interdisciplinary Center for GIS and Spatial Analysis Technologies (CGSAT) and the Graphics, Visualization, and Usability Center (GVU) have combined efforts to take the previously developed software visualization software and revamp it to utilize the state-of-the-art rendering engine from Silicon Graphics through the GL graphics language (Lindstrom et al., 1995). A dynamic quadtree

Figure 4.14. Querying in a perspective image of an integrated GIS

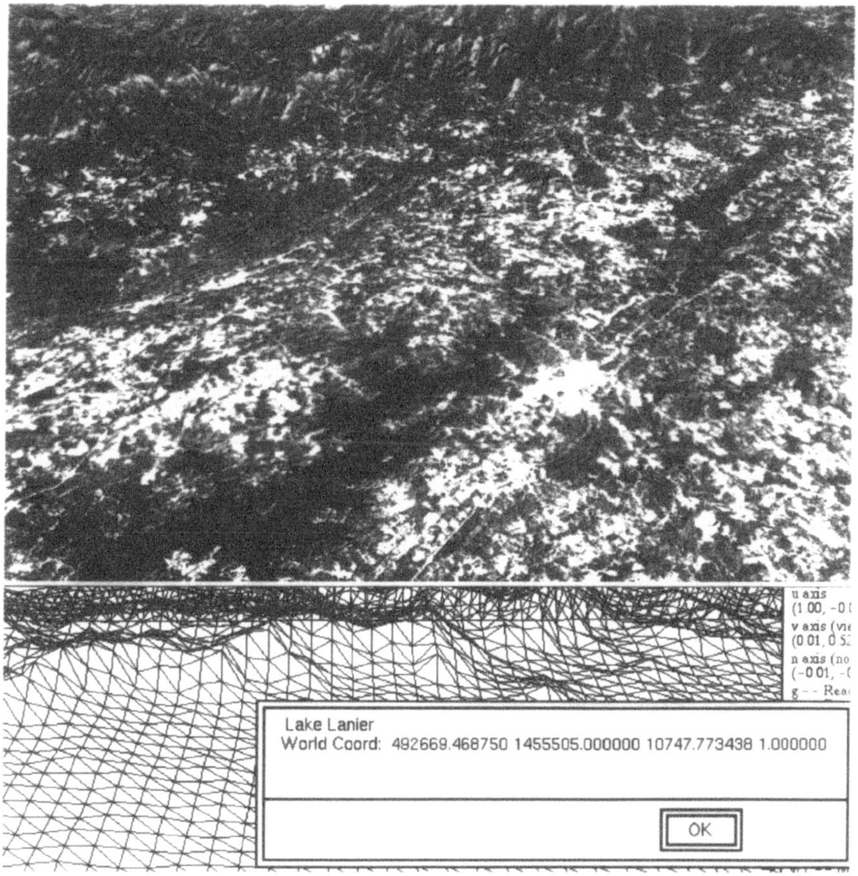

data structure is used in concert with advanced culling techniques to minimize the total number of polygons that must be rendered while retaining the
high resolution through the use of phototexture. The system allows either
workstation window or helmet-mounted goggles to display real-time motion
through a spatial database. The user has the capability of viewing the terrain
from above or to immerse oneself within the terrain. Image and GIS data can
be dynamically switched in real time using a joystick or a data glove from
virtual reality. The user can query terrain or man-made features and read
attributes from a simple raster GIS. Man-made structures are included with
detailed phototexture and intersections with a 3D query results in the name
of the building and other multiple attributes of the building (who owns the
building, the height of the building, the building use, etc.) The user may jump
dynamically from one part of the database to another at any time and will
be able to answer the question of "what can be seen from that ridge line?"
Although the system is still developmental, it shows that the ability to interact
in real time with a virtual 2.5- and 3-dimensional GIS is possible. Several
commercial vendors are investigating the near future addition of a 3D interface to their operational GIS systems. Both high- and medium-resolution
databases have been developed for the virtual reality GIS system. A 1.2-ft
resolution database has been developed for the Georgia Tech campus with
aerial photography, 2-ft resolution elevation data, and phototexture trees and
buildings. A 2-m database has been developed for the downtown Atlanta
area with the major highrise buildings being represented with phototexture.
A 10-m database has been developed for a 50 by 40 km area surrounding
Atlanta, and a 30-m mosaic of *Landsat* Thematic Mapper data is available
for the whole state of Georgia. Logic for smooth transitions between low-,
medium-, and high-resolution datasets is currently being tested within the
software system.

The concept of the virtual GIS described above does not require instantaneous rendering of 3D images. A viewpoint may take a short time (seconds)
to render, but the interactive query, object placement, and general interaction
would happen dynamically. As an extension to this concept, assume the user
has the capability to dynamically traverse the spatial database at any altitude, above or below ground. This Lagrangian motion gives the user totally
free interaction with the database using a data glove, joystick, or other reality
mechanism to provide motion cues. Now a user has the ability to place oneself
on top of a hill and dynamically turn his or her head to see the surrounding
landscape. This capability could be provided using a head or eye-tracker
helmet, which is part of the toolset for virtual reality.

Another technique critical to immersion of the viewer into a virtual spatial environment is the realism of the terrain and objects on the terrain. As discussed above, a rendering system that does not adequately portray terrain roughness or the fine details of a building as it is approached in the virtual fly-by will be seen as cartoonish and will not be accepted as a true representation of reality. By using phototexturing, actual photographs can contribute extensive detail for terrain and a building face. As the user dynamically approaches a building, the model of the building does not break down as a CAD model would. In fact, the closer one gets to the building, the better it looks because it really is a photograph of an existing structure with all the flaws in color and texture that are inherent in real man-made and natural objects. By using phototexture for existing buildings or potential buildings that would have the same architecture, very realistic renderings may be accomplished, allowing users to feel as if they were standing at the point from which the view was rendered. If new buildings are to be visualized that have not been built, CAD models can be imported into the rendering system and drawn into the view.

An optimum virtual reality 3D GIS would also encompass other technologies, such as multimedia. In such a system, a viewer or analyst would be able to move freely throughout the terrain and voxel database; when approaching a man-made existing structure, the viewer may go up to the door and enter the building using real video footage. The system must make a transparent transition between real imagery of the database, phototexture images of the objects, CAD models of unique structures, and the actual video footage.

4.7. Conclusions

This chapter has focused on a definition and extension of the concepts of visualization as applied to the integrated spatial datasets of satellite and aircraft imagery and GIS. We need to consider visualization as not only the tools that have been developed over the years for showing imagery and GIS data, but also new tools that have been developed by the flight simulation, computer graphics, and environmental analysis disciplines. It is now time to investigate an interface into three- and more dimensional datasets with spatial characteristics that will totally immerse a user into the spatial environment, and in fact, this will result in the creation of a virtual 3D GIS.

Acknowledgments

The authors gratefully acknowledge NASA NAGW-1743 and NAGW-3173 for funding during the preparation of this manuscript.

5

Amazonia: *A system for supporting data-intensive modeling*

Terence R. Smith, Jianwen Su, and Amitabh Saran

5.1. Introduction

A *computational modeling system* (CMS) consists of a *computational modeling environment* and *transparent computational support* for the modeling environment. In particular, the computational support may integrate database management tools with an open set of other modeling tools. The goal of a CMS is to increase the efficiency of scientific investigators in the *iterative* process by which scientific models of phenomena are developed and applied. The principal mechanism by which a CMS achieves this goal is through the provision of a unified computational environment in which scientific modeling tasks may be represented in terms of appropriate sets of scientific modeling concepts, and from which a scientist has easy access to a broad range of modeling tools. A CMS can provide much of the functionality required to support the integration of data and software in analyses using remote sensing and GIS techniques.

The modeling environment is based on a characterization of scientific modeling activities that focuses on the manner in which scientific concepts are represented, manipulated, and evaluated. The process of scientific modeling may be viewed as one in which scientists engage in two major sets of activities. In the first set, extensible collections of representations of concepts are constructed, evaluated, and applied in modeling both the phenomena of specific application domains and the phenomena of the modeling process itself. In the second set of activities, instances of the representations are created and sequences of transformations are applied to the instances. In Smith et al. (1994a), we describe a conceptual model that provides a "complete" and consistent foundation for both the modeling environment of a CMS and its associated, high-level *computational modeling language* (CML). The conceptual model is based on a formalization of representations of scientific concepts as *representational structures* (or "R-structures") for concepts.

The main goal of this chapter is to describe key aspects of the computational support for the modeling environment of a CMS. We describe the design and implementation of a distributed CMS prototype, *Amazonia*, and show how it is possible to provide, within a simple, high-level, and unified environment, much of the system support necessary for computational modeling activities, while hiding irrelevant computational details. The contributions of the paper include a hierarchical design for the architecture of a CMS that includes the model environment, data management, and distributed system support; "run-time" support for scientific modeling activities; the uniform integration of heterogeneous tools within a CMS; and distributed operational support. The results reported here grew out of collaborative research efforts with Earth scientists. We are currently applying both the CMS concepts and *Amazonia* to several Earth science problems similar to the example described in Section 5.2.

There are significant differences between our concept of a CMS and the large number of computational systems and modules currently employed in support of modeling activities. Many of these systems provide specialized support for domain-specific modeling activities; others provide tools that support generally applicable but limited subsets of modeling activities. Such diversity of support unnecessarily complicates the modeling process. In relation to the first approach, for example, a variety of distinct computational environments have been developed to support restricted domains of applications without high-level conceptual data modeling, such as *digital terrain modeling* (e.g., Oarg and Harrison, 1990; Kidner et al., 1990; Morris and Flavin, 1990; Weibel and Heller, 1990). Early modeling support systems (Jankowski and ZumBrunnen, 1990; Jones, 1990, 1991, 1993) are specialized and lack support for complex, structured datasets. Computer Aided Software Engineering (CASE) tools and software development environments (Taylor et al., 1988; Liu, 1991; Andrews, 1991; Adams and Solomon, 1993) may also be viewed as specialized modeling environments with respect to software processes (Shy, Taylor, and Osterweil, 1989; Harel et al., 1990; Song and Osterweil, 1994), and many of these tools also manage heterogeneous sets of tools. However, the software development process and scientific modeling possess distinct characteristics: The latter involves large volumes of (scalar) data and collections of algorithms where dependencies (lineage) occur between different projects and teams, whereas the former focuses on computer programs with most dependencies occurring within projects. CMS, in contrast, provides a comprehensive modeling environment based on a general and unifying characterization of scientific modeling activity.

As examples of the second approach, a large number of modules have been developed that support statistical modeling techniques and there have been many extensions of DBMS technology focused on the representation, manipulation, storage, and retrieval of spatial data (Paul et al., 1987; Wolf, 1989; Scholl and Voisard, 1989; Wolf, 1990; van Oosterom and Vijlbrief, Waterfeld and Schek, 1992; 1991; de Hoop and van Oosterom, 1992; Medeiros and Pires, 1994). Recently, there has been an increasing interest in database issues such as heterogeneity, derived data management, and optimization that arise in scientific applications (Long et al., 1992; Chu, 1993). As a result of this interest, several systems have been developed on the basis of relational database technology. In CMS, core concepts underlying the modeling environment are based on significant generalizations of existing constructs in semantic data models (see Hull and King, 1987; Peckham and Maryanski, 1988) and object-oriented data models (see Zdonik and Maier, 1990). In particular, this permits a large range of phenomena to be easily represented at both the conceptual and the implementation level. As a result, it is easy to integrate into CMS many tools that support the modeling environment, including DBMS.

We structure the chapter as follows. We first present a characterization of scientific modeling activity, which we illustrate in terms of an example from the hydrological sciences. We then describe the conceptual basis for the modeling environment of a CMS and provide a brief overview of the computational modeling language CML. The main body of the chapter focuses on a description of the system support for a modeling environment. In particular, we describe the architecture of the system, the modeling support system, the tool management system and its components, the distributed access system, and support for data filtering and datasets.

5.2. Scientific modeling activity

It is critical that computational support for scientific modeling and database activities be based on an appropriate model of such activity. We therefore provide a general characterization of scientific modeling activity and illustrate this characterization in terms of a specific example of modeling activity.

A model of some phenomenon may be characterized in terms of appropriate sets of concepts, representations for the concepts, and applications of

operators to the representations of concept instances, together with an interpretation that maps the symbolic expressions into the domain of application.
In particular, the manipulation of the representations may be characterized
in terms of transformations that map one set of concept representations into
some other set of concept representations. We may therefore view the *process
of modeling* as one in which (1) sets of concepts are chosen or discovered;
(2) representations are chosen or discovered for the concepts, transformations
are defined on the representations, and representations of concept instances
are created; (3) transformations are applied to representations of concept instances; and (4) mappings are made from the symbolic representations into
"phenomena." The phenomena represented in the modeling process include
not only the phenomena of interest but also the process of modeling itself.

5.2.1. An example of modeling activity in hydrology

In order to demonstrate its generality, we illustrate the preceding characterization of the process of modeling, in terms of a specific example drawn from the
hydrological sciences. Figure 5.1 represents a typical sequence of modeling
activities in a modeling process whose goal is to predict the flow of water at
various locations in a river basin.[2] Important classes of modeling activities
illustrated in this figure include (1) the extraction of relevant information from
datasets, (2) the construction and application of models of the flow of water,
(3) the evaluation of the modeled flows, (4) the iterative improvement of the
components of the modeling process, and (5) the communication of the results
of the modeling process.

   In relation to the example, scientific issues concerning the choice or discovery of concepts and representations for the concepts arise in steps 3, 7, 9, 10,
and 14. For example, a question of some interest for a scientific investigator
concerns the manner in which the surface represented in the display of step
2 determines the magnitude and direction of surface runoff. To answer this
question, the investigator may employ concepts that characterize the nature
of the surface and the interactions between the surface and the flow of water.
Examples of such concepts include *surface slope*, *channel segment*, *channel segment contributing area*, and *drainage basin*. If these concepts are to
have relevance in a modeling process in which observations on land surfaces
are represented by digital elevation models (DEMs), it is necessary that they
possess concrete representations that can be extracted from DEMs.

---

[2]The construction of general classes of such models for the Amazon river basin is currently under
investigation by our collaborators at the University of Washington.

There are, however, many modeling situations for which neither a standard set of concepts nor a standard set of representations for concepts has been defined. An important activity in such cases may be the discovery of an appropriate set of concepts and representations. Hence in steps 3–4 of our example, the investigator may (1) create new classes of representations for concepts relating to land surface characterization, (2) extract instances of such representations by transforming DEMs, and (3) evaluate the representations using visual transformations. Similar remarks apply with respect to observations on rainfall events and flow events (steps 7–10). In relation to step 10, for example, an investigator may choose to characterize surface water flows in terms of *surface flow vectors* involving representations of *time*, *location*, and the *magnitude* and *direction* of flows. For the purpose of predicting flows over land surfaces represented in terms of *channel segments* and *channel segment contributing area*, it may be appropriate to construct classes of flow representations in which *surface flow vectors* are decomposed into *channel segment flow vectors* and *overland flow vectors*.

For reasons of convenience, the investigator may wish to employ concepts and representations of concepts that describe *the process of modeling itself*. Such representations could be manipulated during the process of model evaluation and iterative improvement. In particular, once a representation for the entire process has been constructed, it may then be modified, versioned, transformed, and executed as a unit, as in steps 14 and 15 of Figure 5.1.

Issues relating to the definition of transformations and their application to concept representations arise in steps 2–6, 11–13, and 15. In steps 2 and 4, for example, we note that the various procedures on the DEMs may be interpreted as transformations in which one set of representations of land surfaces is mapped into other sets. The *display* command in step 2, in particular, may be viewed as a transformation that maps the representation of a DEM into a representation that is visually meaningful. In step 11, it is necessary to select and apply *solution procedures* to the representation of the flow generating process. The effect of applying such procedures may be interpreted in terms of transformations that map representations of land surfaces, rainstorm events, initial flows over the surfaces, and the flow-generating models into representations of flows over surfaces at various times.

Information may be extracted from such flow representations by the application of further transformations. It may, for example, be desirable to compute representations of *hydrographs* at various locations on *channel segments* (step 12). The visualizations and comparisons of the observed and computed *hydrographs* that occur in the model evaluation step (13) may be interpreted as transformations. The comparison of observed and computed hydrographs may, for example, be viewed as a transformation from pairs of representations of hydrographs into representations of statistical measures.

5.2.2. *Appropriate computational support for modeling activities*

The previous example, with its focus on the manner in which concepts are represented and transformed, illustrates a general requirement that must be placed on computational support for modeling activities. Researchers faced with the task of developing and testing models of complex phenomena in a computational environment must typically work with large numbers of representations of relatively high-level concepts. Such representations may be correctly interpreted only within the particular conceptual framework in which they are used. Hence, providing computational support for such activities requires

Figure 5.1. A hydrological modeling process represented in natural language

1. Get the digital elevation models (DEMs) that intersect the Manaus area.
2. Join this set of DEMs into single DEM and display it.
3. Create a general class of representations for river channel segments and contributing slopes of the drainage system implicit in a DEM.
4. Create and store an instance of this class of representations from the DEM constructed for the Manaus area and store.
5. Find time-slices of rainfall from hour 1 to hour 12 on January 21, 1989, for rainfall records within the Manaus area.
6. Create a rainfall raster for each of the previous datasets using a rainfall interpolation routine.
7. Create a general class of representations for rainstrom events.
8. Find and store all rainstorm events that can be found in the rainfall raster just created.
9. Create a general class of representations for the flow vectors of water on the drainage surface.
10. Construct a model for predicting the surface flow vectors of water over a drainage surface in response to some rainstorm event.
11. Apply the flow prediction model to the specific drainage surface and rainfall event just constructed.
12. Using the map of predicted surface flow vectors, compute the hydrograph that would occur at the downstream end of the highest order channel-segment on the drainage surface.
13. Display a plot of this hydrograph and compare statistically the predicted and observed hydrographs.
14. Create a class of modeling-schemas that encapsulate the sequence of operations in steps 1–13 above.
15. Run this schema iteratively using variants of the flow model from step 10 until the predicted and observed hydrographs are in reasonable agreement.

bridging the gap between these high-level constructs and the low-level details of system implementation.

Traditionally, it has been left to the user to build such a bridge. The procedure is relatively straightforward in conventional applications of relational database technology, since the actual implementation is typically close to the semantics. In advanced applications such as scientific and engineering research, however, the semantics are relatively involved and the conceptual framework is of much greater complexity. It is not reasonable to expect users to make the translation between concepts at the level of the application and concepts supported by the computer. In particular, it would force scientists to focus a major proportion of their effort on issues that are largely irrelevant to their scientific goals.

There is, therefore, a need to support representations of a large and open array of high-level modeling concepts. Some of these concepts relate to the phenomena being modeled, some of them relate to the process of model development, and most of them must be treated as first-class entities. For example, to describe, store, and manipulate information about modeling procedures, a user typically requires high-level descriptions rather than detailed specification of algorithms. The construction, manipulation, and evaluation of models must be described at a level such that the actual variables, routines, procedures, and even datasets used to describe the model are transparent in relation to the main modeling tasks.

It is clear that any reasonable approach to resolving these issues must be based on an appropriate model of scientific modeling activity. The motivation for constructing such a model is to provide a foundation for designing a computational modeling environment in which representations of concepts may be constructed and manipulated in a natural manner. Before discussing the nature of the computational support for such an environment, we first briefly describe the conceptual model that underlies the modeling environment.

5.3. A basis for representing modeling activities

The process by which symbolic models of phenomena are constructed and evaluated may be viewed as one in which a large space of concepts, representations of concepts, and transformations between the representations is explored. Hence, comprehensive computational support for scientific modeling activities may be provided in terms of a computational modeling environment that facilitates such exploration. We have formalized the idea of representation for a concept in terms of representational structures (or "R-structures").

R-structures provide a language for constructing and manipulating representations of concepts and are the foundation on which the computational modeling environment of a CMS is constructed.

R-structures may be constructed that permit scientists to represent, as first-class entities, any reasonable concept that is useful in the modeling process. We may, for example, construct specific R-structures whose R-domains contain representations of primitive abstract entities, such as integers, real numbers, boolean values, and character strings; abstract entities, such as simple and complex geometrical figures; abstract mathematical models, such as partial differential equations; empirically derived entities, such as datasets obtained by various observational means; transformations (e.g., procedures, algorithms, and statistical operations); and scientific modeling schemas that are defined in terms of the representations and transformations of other R-structures. Various "classes" of R-structures capture semantically important distinctions between classes of concepts made by scientists.

In terms of R-structures, we may characterize the process by which scientific representations of phenomena are constructed and evaluated. In particular, we may view scientists as (1) constructing, evaluating and applying collections of R-structures for modeling both the phenomena in specific domains of application and the phenomena of the modeling process itself; and (2) constructing *specific instances* of R-domain elements and applying sequences of specific transformations to sets of instances of R-domain elements. As we note in Section 5.3.2, for example, it is relatively straightforward to translate the hydrological modeling example presented into terms involving an appropriate collection of R-structures.

5.3.1. Representational structures

An R-structure for a concept is a triple $[\mathcal{D}, \mathcal{T}, \mathcal{I}]$ in which (1) $\mathcal{D}$ is the *representational domain (R-domain)* of the R-structure, (2) $\mathcal{T}$ is a set of *transformations* that may be applied to $\mathcal{D}$, and (3) $\mathcal{I}$ is a finite subset of $\mathcal{D}$. The R-domain $\mathcal{D}$ of an R-structure contains a set of representations of all instances of the concept. Since an R-domain will typically contain a large or even infinite number of such representations, an element of $\mathcal{D}$ will typically be specified in terms of a schema that defines the set of representations in some implicit manner. The transformations $\mathcal{T}$ of the R-structure comprise a set of representations of transformations that may be applied to the representations in the R-domain $\mathcal{D}$. These transformations map representations from one set of R-domains to other sets of R-domains. $\mathcal{I}$ is a finite set of representations from the R-domain that are given in *explicit form* and that have particular significance in the modeling enterprise. In general, R-structures are given a name so that they may be referred to as entities.

It is permissible to construct R-structures that possess alternate, equivalent representations of the same concept. In particular, R-structures with abstract representations of concepts (in terms of Object Identifier (OIDs)) are always constructed whenever at least one R-structure with concrete representations is constructed. In essence, a concrete representation is one that may be manipulated with the use of transformations in order to extract useful information.

A simple example that illustrates these ideas involves representations of the concept of *Polygons*. A representation for *Polygons* may be constructed in terms of an R-structure named **Polygons::Points**, where the notation "::" indicates that the abstract domain of polygons has a concrete representation in terms of the domain of Points. In particular, the R-domain of **Polygons::Points** contains *concrete* representations of polygons as sequences of points $(p_1, \ldots, p_n)$ that may be interpreted as the vertices of the polygon. The points $p_i (i = 1, n)$ may be viewed as representations from the R-domain of an R-structure **Points**. In order to make this representation of the concept *Polygons* accord with our intuition concerning polygons, it is necessary to associate with the R-domain of **Polygons::Points** expressions that represent constraints, such as "$p_1 = p_n$" and "*no two edges of the boundary intersect, except at their end points.*" The set of transformations $\mathcal{T}$ on the representations of polygons might include, for example, **Polygons**.*area*, which maps a representation of a polygon into a representation of its *area*, and other transformations such as **Polygons**.*perimeter* and **Polygons**.*centroid*. Finally, we might wish to construct a specific set $\mathcal{I}$ of polygons that are frequently encountered in some modeling activity.

Figure 5.2 illustrates abstract and concrete R-structures for **Polygons**. Further details of the R-domains, transformations, and instances of R-structures are provided in (Smith et al., 1994a). As one might suspect from this figure, various aspects of the R-structure model map *onto* all major aspects of object-oriented models.

5.3.2. R-structures to support modeling activities in specific domains

R-structures appropriate for modeling complex phenomena like hydrological flows in a large drainage basin must include representations for a large set of concepts that range from the primitive to the very complex. To support such applications, various hierarchically structured collections of R-structures may be constructed from some initial set of R-structures. In constructing an appropriate set of R-structures, for some domain of application, one may employ *aggregate constructors* and *constraints*, defined in terms of the transformations, to the elements of previously defined R-domains. An advantage of this inductive approach is that new R-structures may be defined in terms of the sets of concrete representations, the sets of constraints, and the sets of transformations of existing R-structures. This can be done through *inheritance* or *aggregation*. In Figure 5.3 we illustrate a small fragment of a collection of R-structures that possesses an interpretation in terms of polygons.

In relation to the hydrological modeling example, an appropriate set of primitive R-structures includes representations of **Integers**, **Reals**, **Booleans**, and **Strings**. There must also be a collection of low-level R-structures that provide representations of many spatial and temporal concepts upon which modeling is based, such as **Points**, **Line-Segments**, **Polygons**, and **Rectangular-Grids**. There is a need to represent a large body of fundamental concepts,

Figure 5.2. Chart illustrating the concept of R-structures

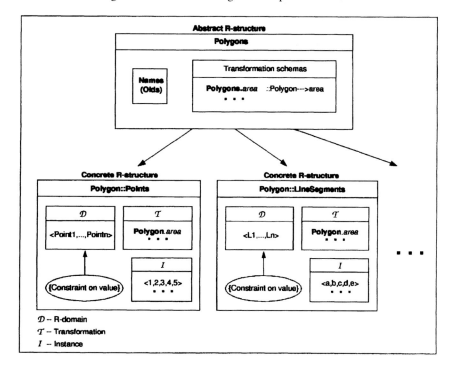

such as **Elevations**, **Flows**, and **Flow-Vectors**, to provide a foundation for more complex concepts. An important class of transformations that may be defined on such fundamental concepts relate, for example, to *dimensional considerations*, such as the need to transform between measurement units.

Based on these low-level R-structures one may construct higher-level R-structures that include various *empirically based* R-structures containing data in the traditional scientific sense and transformations that include statistical operations. Examples of such R-structures include **DEMs**, **Rainfalls**, and **Hydrographs**. Transformations on **DEMs** include those that take a DEM as input and return other representations of land surfaces, such as the channel segments and drainage divides, that may be used in defining the R-domains of R-structures such as **Basins**. One may define still higher-level R-structures that represent relatively abstract concepts, such as **Overland-flow-models**, **Channel-flow-models**, and **Surface-flow-models**. In particular, the R-domain of **Surface-flow-models** belongs to an *implicit* R-structure whose R-domain contains representations of equations that govern such flows and whose transformations include those that solve the flow equations to produce explicit representations of the flows.

In Figure 5.4, we illustrate a small fragment of a collection of R-structures for the hydrologic example. An R-structure from this collection will

Figure 5.3. A fragment of an admissible collection of R-structures

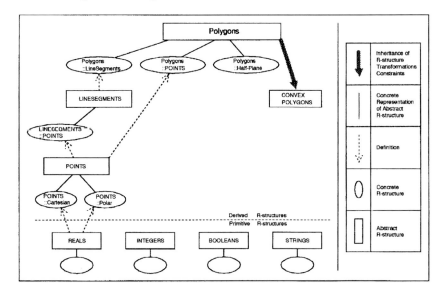

typically have a specification that is similar to the following specification of DEMs:

- **DEMs::peg** represents the concept of *DEMs* in terms of abstract R-structures that include **Points**, **Elevations**, and **Grids** (peg):
- R-domain: consists of a set of tuples of the form $[\{[p, e]\}, g]$, where $p \in$ **Points**, $e \in$ **Elevations**, and $g \in$ **Grids**. The R-structure **Grids** may be defined in terms of parameters that define the geometry of a grid of points. The constraints on these representations include the fact that the finite point set used in defining a DEM forms a rectangular lattice.
- Transformations: *spatial-projection* (maps DEM into its grid of points); *slopes* (computes the surface slope from a DEM); *c-s-extract* (extracts channel-segments from a DEM).
- Instances: *DEMs-Manaus*.

5.3.3. A computational modeling language for CMS

In relation to R-structures as a basis for the modeling environment of a CMS, it is clearly important to have an appropriate language for constructing and manipulating R-structures and their components. We have developed a computational modeling language that may be used to express a large class of modeling activities at the conceptual level of the user (Smith et al., 1994a).

Figure 5.4. A small subset of R-structures for the hydrologic problem

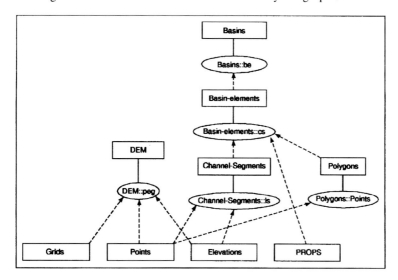

It allows users to express easily and naturally most of the operations that are employed in the iterative development of their models, while hiding computational issues that are irrelevant to the scientific goals.

The primary functionality of CML includes (1) the definition, creation, manipulation, and storage of new R-structures and their constituent parts; (2) the application of transformations to R-domain elements in general and to R-domain instances in particular; and (3) the search for specific R-domain elements and transformations that satisfy appropriate constraints. To provide this functionality, CML possesses a small set of simple commands. This set includes the commands *create, delete, modify, access, store*, which may all be applied to R-structures, R-domains, transformations, and instances, as well as the command *apply*, which applies transformations to R-domain elements. We briefly illustrate the nature of the language with a few examples related to the hydrological modeling example.

The *create* command of CML permits the construction of abstract and concrete R-structures as well as their components. For example, a concrete R-structure DEM may be created using the following command:

```
CREATE DEFAULT CONCRETE R-STRUCTURE DEMs::peg
  SUPER R-STRUCTURES
    = {Rectangular-Grid-Maps::peg}
  R-DOMAIN = [name:string, resolution:integer,
             location: [L1:point, L2:point,
             L3:point, L4:point], P-E:set of
             [Location:point, Elevation:real]]
  CONSTRAINTS = ···
  TRANSFORMATIONS
    = {display-dem(DEMs::peg):bool,...}
```

In this example, the name of the R-structure **DEMs::peg** indicates that the newly created (concrete) R-structure **peg** "implements" the abstract R-structure **DEMs**. If the latter does not already exist in the system, CML will create it with **Rectangular-Grid-Maps** as its super R-structure and with *display-dem* among its transformations.

Transformations and explicit representations of R-domain instances may be created in CML using the *create* command. An advantage of creating transformations within the CML environment is that it is easy to aggregate previously defined transformations into new transformations. This is important in the construction of new R-structures, since the new R-domain elements frequently involve constructors. Hence, transformations that are applicable to these elements will frequently involve, as components, transformations that apply to the R-domains to which the components of the element belong. Each transformation includes a name, a list of parameters and corresponding types, the type of the output value, and a sequence of CML assignments defining the computation.

A key operation in CML is the application of transformations to elements from R-domains. Such applications may be expressed in CML in terms of the *apply* statement. If, for example, the variable Y contains a set of DEM element identifiers, then the command "APPLY DEMs.union TO Y" results in a (new) element of type **DEMs**. The command also returns the identifier of the new element, which can be stored in another variable to be used later. The *apply* command has a large number of important applications, which include the creation of explicit instances of representations (or *datasets* for storage in an R-structure).

CML provides the important *access* command for queries about R-structures, their four main components, and elements of their components. Assume, for example, that the abstract R-structure DEMs has been defined and that *Manaus* is a variable holding a spatial object identifier. Using the predicate *intersect* on pure spatial objects, we can find all DEMs whose spatial projection overlaps the spatial projection of the *Manaus* object using the query:

```
Y = ACCESS{X IN DEMs
            WHERE DEMs.spatial-projection (X)
            INTERSECT DEMs.spatial-projection
            (Manaus)}
```

Finally, we illustrate the use of CML to create transformations that form the basis for representing the concept of a *modeling-schema*. A modeling-schema is intended to represent a sequence of modeling steps that may be accessed by the user in testing, modifying, and applying models. Modeling-schemas possess the characteristics of a transformation, because there are certain inputs (R-domain elements and transformations) and certain outputs (R-domain elements). They may be represented as the R-domain elements of high-level R-structures. Suppose **Spatial** and **Temporal** are two R-structures representing spatial and temporal objects and that the two transformations **Spatial**. *projection* and **Temporal**.*projection* are respectively defined on each. In the following example, we define a transformation **Surface-Flow**.*compute*, which computes a representation of the flow of surface water over some area of interest in terms of a *Hydrograph*. This transformation involves a subset of the modeling steps in Figure 5.1:

```
CREATE TRANSFORMATION Surface-Flow.compute
(o:Spatial, t:Temporal)
RETURN H:Hydrograph
BEGIN
X = ACCESS {D IN DEMs
            WHERE Spatial.projection(D)
            INTERSECT Spatial.projection(o)}
Y = APPLY DEMs.union TO X
Z = APPLY DEMs.drainage-surface TO Y
A = APPLY Rainfall-Map.extract-by-range TO
    Temporal.projection(t),
    Spatial.projection(o)
B = APPLY Rainfall-Map.interpolation TO A
C = APPLY Rainfall-Map.find-storm TO B
D = ACCESS {S IN Surface-Flow
            WHERE(S) INTERSECT
            Spatial.projection(o)
            AND Temporal.projection(S)
            = Temporal.projection(t)}
H = APPLY Surface-Flow.solve-method to Z, C, D
END
```

Once created, such a transformation may be employed in defining the
R-domain of an R-structure for *modeling-schemas.*

5.4. Computational support for a CMS

The second major component of a CMS, and the main focus of this chapter, is computational support for the operations of the modeling environment.
Important aspects of the modeling environment that require support include
the creation, manipulation, and application of R-structures (or the components of R-structures); access to R-structures and their components; and the
application of transformations to R-domain elements. It is also important
that support be provided for the integration of arbitrary sets of computational
modeling tools in a manner that is consistent with the notion of R-structures,
including DBMS as a special case. Furthermore, the computational support
for this functionality must hide scientifically irrelevant details from the user.

We have designed and implemented a CMS, *Amazonia,* that is intended
to provide support for a computational modeling environment based on the

concept of R-structures.   Figure 5.5 shows the high-level architecture of *Amazonia*.

Support for the integrated modeling environment of *Amazonia* includes a CML engine [previously implemented using POSTGRES (Rowe and Stonebraker, 1987) and currently implemented using O_2 (Bancilhon, Delobel, and Kanellakis, 1992)], modeling support, and tool management modules. An icon-based user interface feeds the user commands to the CML engine and the CML engine interprets CML commands. Each abstract or concrete R-structure in CMS is represented by an O_2 class (POSTGRES relation) in the implementation. In particular, an abstract R-structure stores the union of all objects in its corresponding concrete R-structures. The inheritance mechanisms provided by O_2 and POSTGRES are utilized in implementing the sub-R-structure hierarchies. The metadata are also managed using classes (relations). In particular, the metadata include the set of R-structures, transformations, external tools, and structure (type) specifications. The CML engine translates the CML commands into an appropriate sequence of operations in O_2 (POSTGRES) and executes them by directly calling O_2 or the tool management module (for the cases of operations defined in external tools or in POSTGRES, which is treated as a tool). In the latter case, the tool manager executes the corresponding operations through the data access system. The

Figure 5.5. High-level architecture of *Amazonia*

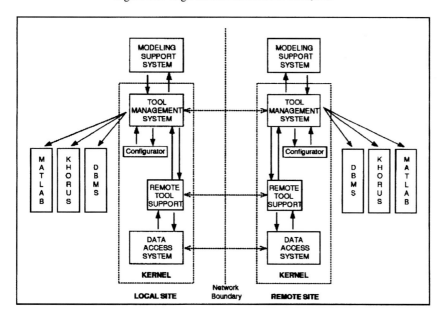

results of the execution are stored in the database, and the information is passed to the user through CML to create the corresponding objects and references at the modeling level.

Scientists access a large variety of modeling tools through the transformations of CML. It is important that support be provided for the integration of this diverse set of tools. For a CMS, integration implies the construction of a generic interface that can handle communication with various tools and, at the same time, keep network and interprocess communication details transparent to the user. The *tool management system* (TMS) provides a single, high-level interface for all external software tools and provides support for such tools in *Amazonia*. The TMS is built over a *distributed access system* (DAS) and employs the services of the DAS to communicate with tools that are locally unavailable. The TMS is designed to keep the system both extensible and flexible and provides an encapsulated environment for accessing any software package. Remote tool support (RTS), as the name implies, provides support for accessing and executing tools that are available remotely.

Amazonia uses a database management system as a tool for persistent storage to support CML. The configurator component of TMS enables *Amazonia* to interoperate with different database systems. The integration of the distributed access and tool management systems forms the *Amazonia kernel*, which provides a unified computational environment that hides *scientifically irrelevant* computational issues from the users. The basic implementation strategy is to tailor modeling environments for specialized applications, such as Earth system science and biomedical engineering, while retaining the underlying integrated tool and data access kernel.

The lowest layer of *Amazonia* consists of the DAS, which provides support for transparent access to data and services distributed over a network. A virtual file system is established for the user in which all data appear local to the user and all data and tool access mechanisms are hidden. The next step in providing computational support is the integration of a diverse set of software tools that might be employed in the construction of models.

We now discuss, in further detail, support for modeling and support for distributed access. We end with a brief description of end-to-end processing in *Amazonia*.

5.4.1. Modeling support module

The *modeling support system* provides support for the modeling environment and, in particular, for model development and model management.

Important examples of model support include the tracking of data interdependencies and providing persistent data structures that represent modeling-schemas. *Amazonia*, therefore, provides capabilities for defining, evaluating, and manipulating complex models in terms of simple manipulations on graphs representing the modeling-schema.

An important task of the modeling support system is to provide persistent storage for model specifications represented in CML as modeling-schemas. This persistent storage of models can be viewed as a traditional database and can be manipulated for querying and composing newer models as well as, for example, maintaining lineage information. As noted above, it provides the functionality to perform a computation in a model as a sequence of execution steps in a single unit.

Internally, each instance (run) of a modeling-schema is represented as a sequence of CML commands that include, in particular, transformations applied to R-domain instances. The implementation employs two types of entities, *objects* and *processes*. Objects are used to represent instances of R-domain elements, whereas processes represent transformations between R-domains. A process takes one or more objects as input and produces one or more objects as output. The exact transformation from inputs to outputs is included as a procedure call that specifies the operation and its execution details. The advantage of this approach is that the user may employ objects and processes to reason about and manipulate complex models. Moreover, to enhance the expressibility of the modeling features, the nesting and functional composition of models is supported.

Information concerning current execution points is stored, and transformations to execute modeling schema instances are thus implemented in a direct way. This provides, for example, the functionality to perform the computations in a modeling-schema as a sequence of execution steps. The user may place breakpoints for checking intermediate results of submodels or processes. Facilities to restart a model are also supported. Other operations such as lineage tracking, consistency checking, and change propagation are also performed at this level. This approach is particularly suitable for cooperative development as has been shown in Alonso and El Abbadi (1994).

The approach employed in *Amazonia* makes maximum use of the concepts of code reusability and version control for modeling as a whole.

5.4.1.1. The tool management system
The tool management system has been developed to *manage* the large variety of tools that are employed in the modeling process. The word "tool" refers to

any executable code, such as a standard software package (e.g., MatLab) or an adhoc program written in an imperative language (e.g., C, Fortran, Pascal). In particular, the TMS supports *transparent access* to these tools. It is critical that such a system be built (1) *without* altering the implementation of tools (since users typically do not have access to the source code), and (2) *without* specific assumptions about any tool (i.e., there exists the capability of adding new tools easily in the existing environment).

In *Amazonia*, tools are accessed through CML constructs, which in turn contain native tool language commands (Smith et al., 1994b). The different ways in which processes may communicate with each other in a Unix environment are characterized by the Unix Interprocess Communication (IPC). A simple way to communicate with a tool in such an environment is to start the tool in an interactive mode and to pass commands using the standard input/output (I/O) redirection facilities available. This approach, however, has several drawbacks: First, it involves significant overhead, since the tool needs to be started every time tool access commands are to be executed; second, it complicates the handling of command responses; and third, it is inefficient due to I/O redirection.

In *Amazonia*, a tool is started only once, it is accessed on an as-needed basis, and we have chosen *pipes* for the purpose of interprocess communication. Although pipes have a standard form, regardless of the Unix implementation, they suffer from inherent limitations that affect much of their usage in the Unix environment. First, pipes support only unidirectional or half-duplex communication between processes. Second, pipes can only be used between processes with a common ancestor. To avoid these limitations, it is necessary to provide additional structures, which in this case involve the concept of Unix coprocesses (Stevens, 1992). Even with coprocesses, however, establishing interprocess communication with tools is a nontrivial task (Saran et al.,1993), because the tools used in modeling are mostly interactive (i.e., they use the standard I/O interface). By default, the standard I/O library uses a buffering mechanism when communicating via pipes. Hence, for correct operation, there is a need to flush the output of the tool to the pipe. This requires a modification in the code of the tool, which is rarely permissible.

To resolve this issue, *Amazonia* supports a generic technique to execute a tool as a background server process. The tool management system uses the concept of pseudo terminals, which ensures that any tool using the standard I/O interface can be integrated into the system (Stevens, 1992). Since a tool runs in noninteractive mode by default, issues such as handling asynchrony of the tool and processing command responses arise. In particular, the TMS must detect command terminations and error sequences. Because these features are tool specific, TMS supports them as configurable parameters specified by the CML in the modeling support system. The TMS permits users to access tools in an interactive mode.

5.4.1.2. TMS components
The tool management system has two conceptually and functionally distinct levels: an upper-level tool box and a lower-level *tool handler* (see Figure 5.6). The tool box handles the processing of language constructs specified by the user that require tool support. It uses the *"channel"* abstraction provided by the tool handler to execute these constructs. The tool handler manages interprocess communication with tools, processing of command responses, and automatic restart of tools and provides the abstraction of channels. A channel is a service access point at which the tool handler services for a particular tool are available. The tool box and tool handler have well-defined interfaces, which are described below.

Since the **tool handler** provides the actual communication to the tool, it uses registration information of a tool, specified at the time the tool is

Figure 5.6. The *Amazonia* kernel

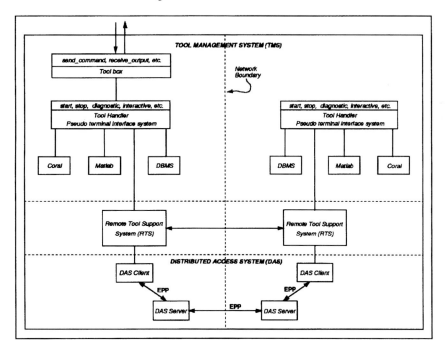

first introduced in *Amazonia*, to initialize and set up the environment for its subsequent execution. A set of commands is provided that may be used for communicating with the external tool:

- *Start* is used to set up the execution environment for the tool and start it subsequently. It returns the channel identifier.
- *Send* is used to send commands on a preassigned channel to a tool.
- *Receive* is used to retrieve the result of the command sent to a tool.
- *Diagnostic* is used to provide information, at the systems level, of the status of the tool process. System crashes and abnormal terminations are also handled as part of this command.
- *Format* is used to take care of any changes in the format of the data retrieved from the receive command.
- *Stop* is used for terminating the execution of a tool in the computational environment.

The above commands are used by the **tool box** to execute user commands. The tool box stores information about each software package, including the name, description, and necessary information and files for execution. Once a tool has been registered in *Amazonia*, this component handles the parsing of internal commands for the tool, processes tool-specific data, and performs language format conversions; it then issues requests to the tool handler for executing the commands. The tool box requests services on a communication link established by the tool handler. As the interface to *Amazonia* kernel, the tool box provides the following simple primitives:

- *Send* is used to send tool-specific commands/transformations for parsing by the tool box and subsequent execution by the tool handler.
- *Receive* is used to collate the output of the command sent to the handler and pass it to the routine invoking the tool box.
- *Interactive* is used for starting an interactive session between the user and the tool, if the user so desires. This mode can be set *on* or *off*.

A TMS at a participating site handles tools locally. If a tool exists at a remote site, the services of the DAS are used. The remote tool support system is used for this purpose. At the local site, remote tools are configured as *Amazonia* data objects. The RTS at the local site communicates with its peer at the remote site, using the services of DAS. The remote RTS in turn makes use of the services of the TMS to access the appropriate tool. Figure 5.6 illustrates the interplay among TMS, RTS, and DAS. Again, the user may wish to run an executable program on some dataset that resides on a remote machine. The TMS, in this case, would invoke the executable program and send a request via the RTS to the DAS to provide access to the remote dataset when needed by the program. Currently, remote data can be accessed on the fly by tools that can be linked with the extended Prospero library, a virtual file system.

The tool management system provides the ability to run tools in the background with the appropriate run-time environment setup. Access to the tools is on a demand-driven basis. We have integrated the front-end tools of DBMS, such as POSTGRES and Coral (Ramakrishnan et al., 1992), and mathematical tools, such as MatLab, with relative ease. An efficient buffering mechanism exists within the TMS to buffer the output of a tool. This relieves the modeling support system from having to determine the size of the output and to allocate space dynamically. Particular attention has been paid to the handling of command terminations and prompts of the tool executing in the background. The TMS can determine the nature of the prompt and its occurrence, to ascertain the termination of the output of the previous command. Because some prompts (e.g., MatLab) have unprintable characters, the TMS interacts with the tool itself to get the prompt by the use of the *command terminator* provided by the user. Another feature of the TMS is the ease with which new tools may be invoked by the user. In traditional systems, accessing a new tool requires low-level knowledge of the tool code or the application programming interface (API). In *Amazonia*, however, adding a tool requires only the specification of a few parameters. Parameters are provided to tune the tool performance.

5.4.2. *Distributed access system*

The DAS forms the lowest layer of the *Amazonia* kernel. It deals with the idiosyncrasies of accessing information scattered across the network. Given the nature and characteristics of spatial data, any realistic approach to the problem of data access precludes a centralized solution. Therefore, DAS is built on a distributed architecture. DAS provides the following three basic features:

- a configurable and uniform interface to heterogeneous data access mechanisms,
- support for remote tool execution and remote filtering, and
- support for proper data abstractions.

The TMS is integrated with the DAS to provide *Amazonia* with a strong kernel that provides support for accessing data and tools, without constraints of network boundaries. *Amazonia* is thus able to establish a comprehensive, consistent computational environment.

Data is often located at different sites on the Internet and possibly scattered around multiple file systems of different types. This makes finding, acquiring, and organizing information of interest a difficult process. There is a need for an abstraction that will hide the heterogeneous file access mechanisms and will allow users to organize information on their own. Although systems like the Alex file system (Cate, 1992), the Jade file system (Rao and Peterson, 1993), Gopher (Obraczka, Danzig, and Li, 1993), and the World-Wide Web (WWW) (Berners-Lee, Cailliau, and Pollerman, 1992) provide facilities to hide heterogeneous access mechanisms, they invariably provide a single hierarchy of organizing information, which does not allow user-centered organization. Hence, we are using Prospero (Neumann, 1992) as a building block for the DAS. The Prospero file system is based on a virtual systems model, which provides user-centered naming, that is, users construct their own virtual systems by selecting and organizing objects of interest, which in our case, include files and directories distributed across the internet. Tools are provided to keep the customized view up-to-date. The Prospero protocol is stateless and is based on the client-server paradigm.

5.4.2.1. *Extensions to Prospero*

The main limitation of the Prospero file system from the point of view of *Amazonia* is that most of its features are restricted to file directories rather than individual files themselves. We have, however, extended Prospero so that it now overcomes the above shortcomings. We use the term DAS server to refer to an extended Prospero server, DAS client, to refer to an extended Prospero client, and EPP to refer to the extended Prospero protocol (Figure 5.6).

We have pushed the actual file accessing functionality from the DAS clients to the DAS servers so that in addition to being directory servers they can also act as data repositories. Thus, local caching is performed at the servers, in cases where the data is fetched from Internet archival sites, which invariably provide anonymous FTP access. No bulk duplication of data is performed. A related issue that arises is that of cache consistency. Since the data being used are typically read-only with periodic but infrequent updates, maintaining cache consistency is not done in real time. Because data are stored in the server, we have extended the protocol between a Prospero client and the Prospero server for data access.

Maintaining a local cache for remote data will not scale well in scientific modeling applications where typical sizes of physical files may be of the order of gigabytes and the data is needed mainly for model testing and evaluation. To obviate the need to maintain local cache, a mechanism that will make remote data available on the fly is necessary. Much of the time, however, the actual data needed in a model execution are only a subset of the available remote data. Thus, it is more appropriate to perform *filtering* of data at the remote site itself, so that only the required data is fetched/transferred. This facility for filtering must also be available in the case of local servers. To perform remote filtering, we have designed and implemented a peer-peer protocol between individual Prospero servers. From the DAS client's point of view, remote tool execution is seen as yet another available access method. We distinguish this access method as PFS (Sastri, 1994).

Services are distributed as a result of the availability of tools and models themselves at various sites, and a user may need to execute tools remotely. Although accesses to services could be implemented through a remote procedure call (RPC)-based protocol, we have chosen an alternative implementation because data and service access will have different interfaces. Rather, we have extended Prospero for remote tool execution, thus providing the user with the same interface for both data and tools. There is no fine distinction between data and service as far as organization is concerned.

Situations might arise when the actual data needed for model execution originates from multiple sources. This is particularly true of scientific modeling applications. Since modeling is a highly iterative process, preparing the data needed for model execution becomes a cumbersome activity if done by the modeler. The DAS supports the high-level notion of a dataset. A user can define a dataset using multiple sources and can access it as needed. To support the dataset and remote filtering, the notions of union and filters in Prospero have been extended to operate at the level of individual files and/or tools.

5.4.2.2. *Support for data filtering*
In Prospero, a filter refers to an executable program that takes a directory as an argument and returns another directory. For efficient data and service access, we have extended the notion of a filter to operate at the level of files and tools (we call such filters data filters). A data filter takes a set of arguments, operates on a single input file or a tool, and generates a filtered output data. To support the concept of data filters, we have introduced an object called "*Amazonia* data object" in Prospero. An *Amazonia* data object is defined in terms of four aspects:

(1) *Data access.* A data source may be a file, in which case this field contains information about file access mechanisms for the data source, or a tool, in which case this field contains information about the tool access mechanisms.

(2) The *filter set* is a collection of data filters that are executed on the data. A filter comes in two flavors: first, as a program that has a single input and

a single output (executable programs will fall into this category); second, as a command script that needs to be executed in a tool environment (database queries and tool access commands will fall into this category). In the latter case the DAS will make use of the services of a TMS.

(3) *Sequencing information* specifies the order of execution of filters. Filters are executed in pipe mode, where the output of one filter is the input to the next filter in the sequence.

(4) Executor information, which specifies the execution environment for a particular filter, for example, shell variables, parameters, etc.

The design and the implementation of the *Amazonia* data object allows one to specify the command line arguments to the filters and to sequence information for filter execution dynamically, as and when the data object is accessed, which will override the default settings. For example, "rain_89" could be a typical data object that refers to the rainfall data for the Amazon river basin for the year 1989, available in the Internet archival site at the Department of Geography, University of Washington (geog.washington.edu) by anonymous FTP, under directory /pub/amazon/data as rainfall.dat. Thus, the data access information refers to file access information.

5.4.2.3. Support for datasets

To support higher level data abstractions at a level appropriate for the user and consistent with the concept of R-structures, we have defined an object called an *Amazonia dataset object*. Such an object extends the concept of union operation on files. This is useful when data from physically distinct multiple sources should be viewed as a single entity. An *Amazonia* dataset object is defined in terms of a *member set*, which is a collection of *Amazonia* data objects, and *ordering information*, which is used to specify the sequence in which the members are to be combined. For example, the rainfall data in the discharge computation problem is usually spatially distributed across different files or as databases in a DBMS. A user can then define an *Amazonia* dataset object for rainfall data, where each member corresponds to a data source, for instance, "rain_dataset" could refer to an *Amazonia* dataset object with a member set consisting of "rain_89" and "rain_90" ("rain_90" is similar to "rain_89", the year being 1990).

5.4.3. An overview of modeling activities in **Amazonia**

We are now in a position to describe the end-to-end functioning of *Amazonia*. The user builds a model of the phenomena of interest, as exemplified by the process illustrated in Figure 5.1, using the visual interface. The user can then execute the model or submit a query based on the model. A query in CML (Smith et al., 1994b), for example, for retrieving the DEMs for the Manaus region, would be:

```
Y = ACCESS {X IN DEMs
            WHERE DEMs.spatial-projection (X)
            INTERSECT DEMs.spatial-projection
            (Manaus)}
```

Upon receiving a command from the user, a translator parses the request to produce a stream of primitive constructs. These constructs are passed to the tool manager, which then interprets them and takes the appropriate action. If the request is to access some dataset, it is passed to the distributed access system, which will determine whether the dataset resides locally or at a remote site, retrieve the information, and pass it back to the tool manager. However, the request may involve the execution of a tool specific command, for example, plotting an elevation map using MatLab, using the CML command "APPLY Display_dem TO Y," where Display_dem is a transformation containing the MatLab commands:

```
CREATE TRANSFORMATION Display (e:DEMs)
    SOFTWARE: MATLAB
    FILE = MatlabFormat(e)
    BEGIN
    load FILE.dat
    plot(FILE)
    END
```

The CML engine sends the request for execution of the transformation to the TMS. On receiving the *send_command*, the tool box accesses the transformation from the library. After associating the positional parameter (the OID of the FILE to be displayed), it ascertains the software to be MatLab and retrieves the initialization and execution information for MatLab. This information is then used in the *start* command of the underlying tool handler, to execute MatLab. *Start* returns the channel link information. The tool box then proceeds with a format translation of the hydrograph input that is understandable by MatLab. Using the *send* primitives, the commands (*load* and *plot*) are sent to the background process MatLab. Subsequently, the output of the commands are received in order, by a sequence of *receive* primitives.

The tool box manages a buffer pool, which stores the output of the tool. This output is passed to the invoking process (in this case the CML engine), depending on the size requested in the *receive_output* command. The hydrograph is plotted by MatLab on a new window. The *diagnostic* primitive is used in case alternative action has to be taken for abnormal execution. Note that for subsequent invocations of MatLab in the same environment, a new process is not started. The tool handler checks to see if it is already running, and if so, passes the link information to the tool box when the *start* command is used. In this case the tool manager must determine in which execution environment (MatLab, in the current example) this is to take place. If the tool to be used is local, the local tool manager will handle the execution and open a new window to display the map. If the tool is remote, the tool manager will start a protocol with the RTS to arrange the execution of the command. This system communicates with the DAS, which sets up the link with the remote site.

5.5. Conclusions

The concept of *computational modeling systems* is of value in providing natural and transparent computational support for scientific modeling activities. This conclusion is based to a significant degree on the results of close research interactions with Earth scientists. In particular, we conclude that it is important that such systems be based on a foundation that provides a complete and consistent characterization of the process of modeling. The concept of *representational structures* not only provides such a foundation but also brings a unifying simplicity to the computational modeling environment of a CMS, to the associated high-level computational modeling language, and to the integration of computational support modules.

Amazonia is a CMS that has been constructed on the basis of these principles. It provides a coordinated set of utilities and tools that cooperate under a unified environment and that are integrated within a single system. *Amazonia* has two major components: a *computational modeling environment* that is designed on the basis of the model of scientific modeling activity and *computational support* for the modeling environment. In relation to the computational support component, the *modeling support system* of *Amazonia* provides support for both model development and model management, as well as persistent storage for model specifications represented in CML as modeling-schemas. The *tool management system* of *Amazonia* has been developed to manage the large variety of tools employed in the modeling process. In particular, the TMS supports *transparent access* to these tools and includes support for scalability and extensibility. The *distributed access system* of *Amazonia* is built on Prospero and provides support for the organization of data and services, both local and remote. It hides heterogeneous access mechanisms through a single interface. For efficient data access to large objects, remote filtering and remote tool execution are provided. Facilities are provided to make the data available on a demand-driven basis.

5.6. Acknowledgments

This work is supported by NSF grant IRI-9117094 and NASA grant NAGW-3888. The authors wish to thank Amr El Abbadi, Divyakant Agrawal, and other members of the Amazon project team for their contributions.

6

Accuracy

Kenneth McGwire and Michael Goodchild

6.1. Introduction

As the fields of remote sensing and GIS mature, these two disciplines are increasingly being looked to for ways of quantifying complex social and environmental events. Decision making regarding such events generally involves uncertainty in understanding, quantification, and prediction. In order to assess alternatives in a reasonable manner, some indication of confidence in various information sources must be available. This chapter identifies aspects of spatial data processing that affect the accuracy of information products derived from remote sensing and GIS data analysis. The discussion is set in the context of a spatial data processing strategy for providing information products of high quality and known accuracy characteristics to decision makers.

Though the integration of remote sensing and GIS has been promoted for some time (Shelton and Estes, 1981, Marble and Peuquet, 1983), and to greater or lesser extents implemented, only recently has there been some recognition of the need to draw from the background of both communities to synthesize an integrated view of accuracy for spatial information (e.g., Chrisman, 1989). Error accumulation in remote sensing and GIS data processing is difficult to track, both in terms of the availability of data for validation and the conceptual understanding of error sources, their propagation, and individual or cumulative effects. In remote sensing, the acquisition of consistent spectral response from the Earth's surface is made difficult by the complexities of sun/target/sensor geometry, bidirectional reflectance distributions, nonuniform atmospheric characteristics, and spectral variability (Duggin, 1985; Asrar, 1989). In GIS, positional and thematic accuracies are a compromise of scale, contemporaneity, media stability, and compilation standards (Mahling, 1989; Goodchild and Gopal, 1989). As will be shown, many of the uncertainties encountered within these two complementary disciplines are analogous. The challenges of representing social and environmental phenomena accurately with remote sensing and GIS will be considered as falling within a more general heading of spatial data processing. The following chapter sections organize accuracy issues within spatial data processing into five interrelated areas referred to as process, measurement, format, analysis, and assessment. Figure 6.1 presents the framework for this discussion.

Figure 6.1 poses a question regarding a spatially distributed process and then develops a strategy to determine functional relationships between observations. This strategy defines the methods that will be used to transform available data sources into information products for decision making. The ability to provide accurate information regarding the spatial process will be fundamentally dependent on the level of understanding of the system under study. Issues related to the original delineation and description of a system with regards to a particular query will be discussed as PROCESS. Data collected to quantify components of the system under study are subject to errors or uncertainty in MEASUREMENT. Data FORMAT, which is required to establish a consistent and manageable data structure, may also limit accurate representation of natural or social measurements. ANALYSIS of data may use

Figure 6.1. Accuracy issues in spatial data processing

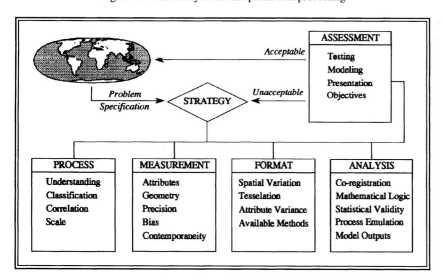

some accepted relationship between process components or may develop new relationships between data sources. The techniques used to synthesize new information products may strongly affect accuracy. Finally, prudent experimental design also requires that direct accuracy ASSESSMENT be performed prior to acceptance of results. If the observed system is not adequately represented by an information product, then the initial strategy must be examined for flaws, and a new strategy may be formulated.

The strategy selected to address a particular question defines the types of data used, the methods of analysis, and the ability to assess results objectively. This strategy will be affected by numerous indirect considerations, including time, cost, and expected short-term versus long-term benefits. Specific strategies will vary by application and circumstance; however, all spatial data processing scenarios will require consideration of some, if not all, of the generic concerns outlined in this chapter.

6.2. Process

The data and operations of a GIS represent a workable model of a physical and/or cultural system. To answer a query regarding a particular system, the relevant observations with which to represent that system must first be identified. In order to obtain relevant data to answer a query, some level of understanding is required regarding the process under study. This knowledge allows adequate delineation of relevant boundaries of a system and determination of the most parsimonious solution to a query. Examples of how inadequate knowledge of process may reduce accuracy include the use of inappropriate generalizations, mistaking causation from among correlated variables, or focusing on an inappropriate scale of observation.

Classification of phenomena within a discrete taxonomic scheme may facilitate identification and communication regarding complex phenomena. Inherent in any classification scheme is the concept that within-class variability is less than between-class variability. In a spatial context, this principle applies not only to the definition of the classes, but also to the regions created when the phenomenon is mapped. Despite this, the way in which features are classified may vary depending on the goals for which the taxonomy was developed. To ensure accurate representation of a phenomenon, it is important to understand the context under which original categorizations were developed prior to their acceptance in subsequent studies. The classification scheme developed for image analysis by Anderson et al. (1976) may be used to demonstrate this situation. This taxonomy was originally developed in the eastern United States for mapping land use and land cover. Though intended to be general and flexible, this scheme may not be optimal for general studies of natural vegetation. For example, chaparral vegetation of Mediterranean climates might be classified "rangeland" at the coarsest level of the taxonomy, even though no grazing is possible, the above-ground biomass may be similar to forests, and the community has very unique compositional and structural properties.

In addition to understanding the generalizing characteristics of a chosen observation, interrelationships in the system under study must also be understood. GIS-based analysis has increased the number of social and environmental variables that may be cross-referenced to answer a query. As with the problem of multicollinearity in statistics (Montgomery and Peck, 1982), there is the potential to mistake the correct source of causation from among a number of correlated variables in GIS-based analysis. Thus, approaching spatial data analysis without sufficient understanding of the subject under study may result in inaccurate inference. The use of correlative relationships may limit the accuracy and extensibility of observations used to address a query. The variable accuracy of automated classifications derived from remotely sensed data exemplifies this general problem of correlative relationships. Spectral reflectance characteristics of the landscape may be correlated with land use/land cover patterns, but specific spectral reflectance patterns are not always inherently tied to classes with informational value. Thus, automated identification of certain classes may be difficult, and extensions of locally generated relationships between spectral reflectance and land use may break down over time and/or distance. Classification schemes that causally link information classes to spectral response have provided some improvement in accuracy (Jensen, 1978; Running, Loveland, and Pierce, 1994), although the information content of these classes may be specific to certain applications.

The scale of observation used to characterize a system may also have a profound effect on the accuracy of information derived from spatial data processing. In answering a query, the extent and detail of observation required to delineate a system in time and space must be determined. The resulting spatial and temporal scales of observation establish a limited frame of reference regarding the process under study. Both data and analytical methods may be tied to specific, but potentially conflicting, frames of reference. Observed patterns in a system may vary widely depending on the degree of precision and range of conditions under which observations are made (Getis and Franklin, 1987; Turner et al., 1989a; Moore and Keddy, 1989). This situation has been widely documented from parametrization of evapotranspiration (Jarvis and McNaughton, 1986) to the modifiable unit area problem of spatial econometrics (Openshaw and Taylor, 1981). Several papers have addressed such scale-dependent patterns in both GIS and remote sensing applications

(Turner et al., 1989b; Townshend and Justice, 1990; Stoms, 1992; McGwire, Friedl, and Estes, 1993).

The analytical methods or process models that use spatial data may also be limited to the frame of reference under which they were conceived. This may arise when the driving processes of a system are dependent on the historical state of that system. For example, the presence of a certain forest species in an area might be more dependent on the occurrence of a recent disturbance (fire, landslide, etc.) than it is on temperature and moisture regimes. Though one might be able to characterize the climatic conditions that allow a single tree to grow at the current time, an understanding of the factors affecting the presence of the species in that region would typically require analysis at broader spatial and temporal scales.

The spatial and temporal scales selected to represent a system must be compatible with both the scope of required information and the expected range of system variability. The graph in Figure 6.2 identifies the approximate scales at which selected applications (horizontal entries) and data sources

Figure 6.2. Time/space scales of applications and data sources

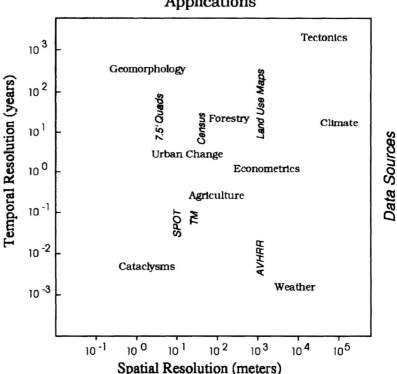

(vertical entries) occur. Many methods have been used to quantitatively describe the scale-dependent behavior of spatial data, including fractal, geostatistical, block variance, and power spectrum techniques (Ludwig, 1979; Lovejoy and Schertzer, 1988; Legendre and Fortin, 1989). These methods may be useful in determining whether the characteristic variability in a dataset is compatible with the temporal and spatial scales of process models and subsequent decision making.

6.3. Measurement

Once the general types and resolutions of observations that are needed to quantify a system are determined, the specific nature of available measures must be understood. Acquiring and entering data often comprise a large proportion of the total cost of a GIS (Kennedy and Guinn, 1975). Additionally, many environmental and social variables cannot be exhaustively quantified. The technical limitations of data collection, storage, and processing will affect the ability to quantify spatial phenomena accurately. As a result, the data products used in spatial analysis are often surrogates for desired observations, being selected on the basis of availability as much as their functional significance in the system under examination. At a more fundamental level, all measurements are imperfect or incomplete representations of a natural system. The ability to understand the accuracy characteristics of derived information products depends on understanding the characteristics of available data inputs. Although the philosophical basis of measurement and sampling theory is beyond the scope of this paper, accuracy issues arising from the precision and bias of measurements will be discussed in this section.

6.3.1. Precision

Whereas accuracy is defined as a measure of the difference between a measured value and the truth (often defined with reference to a source of assumed higher accuracy), precision is defined as the degree of detail in the reporting of a measurement and is most often determined by the characteristics of the measuring instrument. In principle, measurements should be reported with a precision that matches their accuracy, but that principle is often ignored in data processing applications when results are reported to the maximum precision available to the system. Products with a relatively low degree of thematic or spatial precision may still be useful so long as the resulting uncertainty is matched to the requirements of the original query. For remotely sensed data, spatial resolution will often be identified by pixel size or instantaneous field of view. In cartography, accuracy has typically been addressed through the map scale (in this context, a representative fraction) and the existence of a variety of map compilation standards. As an example, the National Map Accuracy Standard (NMAS) of 1947 historically used by the U.S. Geological Survey requires that 90% of samples selected from well-defined points are within 0.025 inches (0.64 mm) of their correct positions. Thus, a 1:24,000 scale map might be assumed to have a spatial resolution of about 15 m. However, the assumptions that relate map scale to spatial resolution may be misleading, especially since many land cover maps show only those features that exceed a larger threshold size, termed the minimum mapping unit (MMU). Thus, for some maps the positional accuracy of polygon boundaries may be less useful as an indicator of spatial precision than the MMU, particularly with regard to land classification.

In the context of GIS, the predominant form of measurement occurs in the digitizing or scanning of maps. Given the traditional reliance of GIS on inputs from analog maps, much attention has been focused on uncertainty in the creation, digital representation, and registration of feature boundaries. One of the classic monographs in this area by Peucker (1976) addresses the fundamental nature of the cartographic line and relates the accuracy of digitizing to the rate of sampling with respect to the curvature of boundaries. The variability between operators in digitizing selected features has also been studied empirically (Maffini, Arno, and Bitterlich, 1989). As a result of variability in data capture, the spatial resolution of data obtained from a map may be somewhat coarser than that of the original product. However, the various stages of digitizing or scanning rarely introduce positional uncertainties of more than the 0.5 mm already present in most map products and allowable under traditional map accuracy standards.

In remote sensing, the spatial, spectral, and radiometric resolution of a sensor determines the accuracy with which features may be measured. Positional accuracies for the *Landsat* Thematic Mapper and *SPOT* sensors have been likened to those of 1:50,000 and 1:25,000 scale maps, or 32 m and 15 m, respectively (Welch, Jordan, and Ehlers, 1985; Konecny et al., 1987). The geometric fidelity of digital imagery is affected by both sensor and platform characteristics. Though airborne scanning systems are capable of higher spatial resolution than satellite-based platforms, aircraft platforms are subject to greater variability in platform attitude and altitude. Their proximity to the ground also increases topographic distortions and off-nadir scan angles. A number of sophisticated efforts at correction of airborne scanner data are

being developed, including real-time GPS measurements and more sophisticated correction algorithms (Fisher, 1991; Ehlers and Fuller, 1991; Fogel and
Tinney, 1994). These issues are addressed in more detail in Chapter 2 of this
monograph.

The spatial accuracy of satellite-based image data continues to increase
with both improved spatial resolutions and wider availability of precision
topographic correction. This increase in spatial resolution combined with
pointable sensor design will also allow topographic mapping in remote areas at a much higher level of accuracy than may be found in existing map
products. Though higher spatial resolution tends to increase the information
content of image data, the associated increase in scene complexity may make
automated classification of certain surface features more difficult (e.g., Toll,
1984; Williamson, 1989). In contrast to increased spatial resolution, sensor
design in the coming era of remote sensing for global change research is focusing on the development of more precise spectral measurements at coarse
spatial resolutions, which will facilitate global data coverage (e.g., MODIS-N
has 36 spectral channels).

In addition to accuracy limitations in manual data capture or direct measurement, the accuracy of data sources may be altered by spatial processing.
As previously discussed, the validity of attribute measurements may be tied
to the spatial resolution of original observations. Processing may increase or
decrease the spatial precision of derived data products in a way that may lead
to inaccurate interpretation. In one sense, features in data products may be
represented with spurious precision. Such a situation may occur in the vector
domain through compositing of data with differing spatial resolutions into a
single data product. In the raster domain, data may be resampled to finer resolutions using interpolation criteria that do not take into account the underlying
distribution of the phenomenon being measured. For example, cubic convolution is often used to resample image data because the resulting product is
visually more appealing than alternate methods. However, resulting data values may exceed the actual range of reflectance. Similarly, the new generation
of digital elevation models being generated by the U.S. Geological Survey are
interpolated from hypsometric data in 7.5′ quadrangles. Although the models provide a better behaved surface than the previous Gestalt photomapping
methods, this interpolation uses limited information on the nature of real topographic surfaces and its assumptions may be naive. More specifically, the
interpolation uses a fixed weighting function rather than adapting to local
patterns of variance.

Alternatively, processing may degrade the spatial precision of data products by convolving values found within a neighborhood. Examples of such operations include calculations of slope and aspect from digital elevation models or low-pass filtering of digital image data. Typically, algorithms that calculate slope and aspect do not resample output values to the neighborhood over which they are representative. Using these derived products at the spatial precision of source data may cause inaccurate interpretation, especially if the scale of the original observations was marginal for a particular application. In some cases, the Nyquist sampling theorem may provide guidance in determining the limitations of spatially convolved data products.

6.3.2. Bias

A second accuracy issue is whether measurements display bias with respect to attribute or position. Whereas random fluctuations may cause imprecision in measurement, bias refers to systematic differences in measurement characteristics, which may also depend on the location or time of measurement. Bias may range from a lack of completeness in enumeration to an inconsistent relationship between a phenomenon and the measurement technique being used. Information products derived from biased measurements may misrepresent relationships in a system if the bias is not recognized (for further details, see Section 6.4). The simplest case of attribute bias is that of miscalibration – a systematic inconsistency regardless of location and land cover type. For example, spectral measurements in remotely sensed data may be systematically misrepresentative of surface reflectance if sensor calibrations and atmospheric corrections are not properly applied. In GIS analysis, simple bias in positional accuracy may arise when data are digitized from map media with unstable shrink/swell characteristics. Changes of up to 2% in the dimensions of paper map products have been documented with extremes in temperature and humidity (Monkhouse and Wilkinson, 1973). Failure to recognize differing datums in map sources may also cause simple positional bias (e.g., NAD27 vs. NAD83).

When bias in a data product is dependent on the feature, location, or time of measurement, it may become difficult to state in what ways resulting information products are affected. In such a case, the risk of using information products may vary based on the alternatives being considered. Complex biases may affect both the attribute and position of mapped features. An example of feature-dependent, attribute bias occurs in map products when uniformity of detail is sacrificed for selective representation of those features that are useful for orientation (Roth, 1991). Interestingly, the 1947 NMAS does not address the thematic accuracy of mapped features. Similar feature-dependent, thematic bias is encountered in automated land use/land cover classifications derived from remotely sensed imagery. Because relationships between an object's spectral characteristics and its rank and associations in a land use/land cover classification scheme may be weak, errors in such machine-generated land cover maps are generally not consistent across classes. As a result, the overall map accuracy statistic may not be relevant to a particular application. Though the Anderson classification scheme (Anderson et al., 1976) developed for image interpretation addresses thematic bias by specifying that accuracy "should be about equal" for all map classes, this criteria acts only as a guideline for product compilation and does not provide a basis for conceptual error modeling.

Attribute bias may also be dependent on the size of the feature being measured. In remote sensing, the spectral response of surfaces will be blurred by both the intervening atmosphere and sensor optics. As a result, spectral measurements for smaller features may not be as accurate as for features covering large fields of view (Kaufman and Fraser, 1984). Similarly, in the GIS domain, Turner et al. (1989b) document the preferential deletion of specific land covers with small spatial extent as spatial precision decreases. Attribute bias may be location dependent as well, as in the case of uncorrected atmospheric effects that vary within a remotely sensed image.

Positional bias may also vary in a complex manner depending on location. An example of variable positional bias can be observed in image data that has not been corrected for topographic distortions. These distortions are localized scale changes and occur as a result of varying distance between the imaging system and the land surface (Paine, 1981). Complex positional bias may also be introduced to planimetrically accurate source data through nonlinear coordinate transformation. For example, confusion regarding map projection parameters may create distortions that surpass simple coordinate offset.

6.3.3. Temporal issues

Though the temporal dimension is usually fixed in spatial data products, issues of measurement precision and bias are still relevant. Efforts are currently being directed at the compositing of global or regional data sets to support global change research (e.g., IGBP, 1992). Such composite data must capture a consistent view of the landscape despite dynamic surface and atmospheric processes that change during and between the times of image acquisition. Despite such time-dependent variations, the high temporal repeat rate of the Advanced

Very High Resolution Radiometer (AVHRR) sensor allows composites of this data to correspond well with certain environmental parameters (Tucker et al., 1983; Prince and Tucker, 1986). Temporal inaccuracy in GIS-based analysis may also result from a lack of database concurrency, since a GIS database is a static representation of what is often a dynamic system. Not only might the database misrepresent the current status of a natural system, but data acquired from different periods may create a representation of states that never exist contemporaneously. Such concurrency issues suggest that a method of assigning lifetimes to data products should be developed. However, the enforcement of temporal validity may be intractable due to the unpredictable or discontinuous nature of certain processes. The issue of database concurrency is a prime motivation for more sophisticated integration of remote sensing and GIS technologies.

6.4. Format

Digital analysis may allow more rapid and flexible assessment of complex systems than manual methods. To manipulate digital measurements effectively, some common framework for data analysis must be adopted. The selection of a particular database structure, or data model, may significantly affect one's ability to pose queries and derive information (Date, 1986). In GIS, the choice of a data model affects the inherent capacity both to represent spatial phenomena and to characterize product accuracy. Ideally, the choice of a data model should be driven by the need for accurate representation of real spatial variation, in order that decisions based on GIS analysis be as reliable as possible. In reality, the choice of a data model is often driven by the limited capabilities of particular software choices, by the constraints of measurement systems, or by the user's experiences and biases.

Measurements of simple scalar values, such as the distance between two points or the height of a tree, are easily represented as numbers and are readily transferred to the digital environment. However, the digital representation of spatial variation requires much more sophisticated approaches. Spatial data handling systems provide a variety of data models for defining attributes within a two-dimensional field (Goodchild, 1992). In the context of remote sensing–GIS integration the user is frequently limited to two data models, termed here as the raster and polygon models. Each data model has advantages and disadvantages in terms of maximizing and quantifying the accuracy of spatial representations. The raster model quantifies landscape attributes within the explicit control of a systematic sampling frame. The term "polygon model" is used in this chapter in place of the more common term "vector" in order to clarify certain functional requirements of a data model for GIS analysis. We define the polygon model as a piecewise approximation to a two-dimensional field, in which the plane is divided into nonoverlapping and arbitrarily shaped regions based on the attribute being represented. This is contrasted with the limited representational capabilities of vector-based computer aided drafting (CAD) packages, which also build polygons from line segments, and line segments from points, but do not enforce consistent organization of features within a two-dimensional field.

To maximize the accuracy of a spatial representation, the choice of either data structure is dependent on the features being sampled. The polygon data model is generally used to represent various types of area classification, or area class maps, for landscape attributes such as soils, land use, or land cover. Although allowing precise measurement of the positions of polygon boundaries, the geometric precision of the polygon model may have little to do with accuracy. Mapped boundaries are often no more than crude approximations to broad zones of transition, and the polygons they define are often far from homogeneous. The potential of the polygon model for greater positional accuracy is justified for cases where changes in land characteristics occur along well-defined boundaries between relatively homogenous spatial units. Those features which are continuous, or for which the scale of observation makes precise taxonomic determination impossible, might be quantified most effectively within the systematic sampling of the raster model. As an example of the former case, Kumler (1992) documents greater accuracy in representing continuous topographic surfaces with raster digital elevation models than with the vector encoding of a triangulated irregular network (TIN). Certain data sets used in general circulation models (GCMs) represent the latter case by coding land use/land cover as percentages of primary and secondary classes within a raster. The spatial control of the raster data structure makes systematic determination of percent cover for each land cover type far more tractable than would be possible in the vector domain. However, the traditional, rectilinear sampling of the raster model may create distorted representations of spatial variation as well. For example, to reduce problems with directional bias found in the typical raster model, Burroughs (1988) utilized a hexagonal tessellation for a fire simulation. The challenges of global scale studies may also exceed the representational capabilities of simple, raster data structures. In response to this, Goodchild and Yang (1992) have developed a tessellation of the globe based on triangular decomposition of an octahedron for which facets have approximately equal area and shape.

In an integrated remote sensing and GIS environment, both data models must often be used in conjunction. The pixels of a raster data file act as the spatial objects whose numerical attributes form an approximation of the two-dimensional field being measured. In some applications, such as land cover classification of image data, these attribute values may be used to classify each pixel into one of several classes. Contiguous pixels with identical attributes may then be grouped to form zones of uniform class, and the boundary of each zone may be identified as an ordered set of coordinate pairs in some suitable coordinate system. This process produces the polygon model, as spatial variation is now described by a partitioning of the space into irregularly shaped polygons. Although each polygon will be homogeneous with respect to class, there will likely be substantial within-polygon variation in spectral response. This will be especially true if the polygons are subsequently aggregated in order to generalize the representation to coarser spatial or taxonomic resolution. However, if raster-derived polygons are not suitably generalized before converting to the polygon model, topological relationships may become confused along complex polygon boundaries. Alternately, conversion from the polygon to raster model also introduces uncertainty with respect to the position of feature boundaries. Frolov and Maling (1969) use the size distribution of cells bisected by line features to provide an error estimate for this effect. Goodchild (1980) later refined this estimate by introducing the effect of serial correlation in line segments.

The chosen data model will also affect the ability to characterize uncertainty in data products. Perhaps the most commonly cited method for characterizing uncertainty of boundaries in the polygon data model is the epsilon band (Perkal, 1956, 1966; Blakemore, 1984). The epsilon band is a zone around the observed position of a line within which the true position of the line is expected to lie with some measure of confidence. Some sources of error, such as digitizing, may contribute a constant epsilon. However, for many applications of GIS to land classification, the variable width of transition zones between adjacent polygons cannot be adequately represented by a constant epsilon distance. Although the epsilon band concept is not necessarily limited to the polygon data model, its implementation in a raster approach may be inefficient. This inefficiency arises because topological information is not explicitly stored to indicate which neighboring classes might be confused, and the spatial resolution required for characterizing boundary uncertainty may be more detailed than that of the original raster data file. Although the epsilon band provides a useful way of describing uncertainty in a line's position in the polygon model, it has not been possible to formalize it as a parameter of a statistical model of uncertainty or to make rigorous connections between it and those statistical models that have appeared in the literature (Keefer, Smith, and Gregoire, 1988; Goodchild, Sun, and Yang, 1992). Thus, the epsilon band remains a useful, but isolated, concept.

In general, not all points contained within a polygon will be correctly represented by the assigned class. Spatial information on such within-class variance may be more easily represented with the raster data structure. Continuous error estimates within the field of measurement may be possible for both categorical and continuous raster attributes. Field-based error modeling within categorical data products is typified by the per-pixel confidence values derived from maximum likelihood classification of land cover in remotely sensed image data. The kriging interpolation technique provides an example of field-based error modeling for continuous, spatially autocorrelated measures. Kriging uses an empirical model of spatial autocorrelation to create error estimates for every interpolated point (Journel, 1989). As with the polygon data model, attribute heterogeneity occurring within raster cells will generally be unavoidable. The kriging approach may be implemented using a block method to appropriately estimate error variance within a grid cell, rather than for a specific point. Unfortunately, it has not been possible to make analytic connections between the field-based view of uncertainty and that inherent in cartographically based descriptors, as noted previously with the epsilon band.

6.5. Analysis

The topological foundation of GIS permits assessment of spatial relationships beyond the abilities of computer-aided drafting and relational database approaches. GIS permits a holistic approach to system characterization that allows unique information to be synthesized from disparate data sources. However, the ability to accurately integrate multiple data sources is first dependent on the degree to which absolute geometric registration between data sources can be enforced. Inconsistencies in boundary location between data sources result in the creation of spurious, sliver polygons during intercomparisons. Boundary uncertainty between data products is pervasive, as perfect repeatability is not possible in either map compilation or in the digitizing process. Specific problems arising from cartographic overlay have received much attention (Mead, 1982; Newcomer and Szajgin, 1984; Veregin, 1989; Chrisman, 1989). Goodchild and Gopal (1989) ascribe this difficulty in computerized spatial representation to an inability to compensate adequately for differences in map characteristics in a manner that is comparable to manual

Table 6.1. *Mathematical operations associated with measurement types*

Measurement	Characteristics	Examples	Valid Operations
Categorical	Classification into a taxonomy where the ordering of class values is arbitrary.	Soil series Political jurisdiction Acceptable/Unacceptable	Equals
Ordinal	Relative ordering made with unknown or unequal intervals between groupings.	Highest → Lowest Fastest → Slowest Standard of living	Less than Greater than
Interval	Continuous measurement using equal intervals made from an arbitrary zero point.	Sensor DN value Elevation above sea level Degrees Fahrenheit	Addition Subtraction Scaling by a constant
Ratio	Continuous measurement using equal intervals made relative to an absolute value of zero.	Calibrated irradiance Income Degrees Kelvin	Multiplication Division

interpretation. Misregistered features not only change positional or area estimates in resultant information products, but may also generate inappropriate
relationships between landscape features. In this latter case, derived information products may inaccurately represent the coexistence of environmental
or cultural features of the landscape. As mentioned in the previous section,
the epsilon band approach may allow for automated resolution of boundary
uncertainty.

The accuracy of spatial data analysis is also dependent on appropriate application of mathematical, statistical, and process-emulating manipulations.
An example of limitations in mathematical manipulations may arise in the
common GIS analysis of generating a weighted, linear combination of data
products. This approach may be used in site suitability scoring or in generating
indices such as the universal soil loss equation (USLE). Commercial GIS systems do not currently have the metadata management capabilities to enforce
mathematical logic in overlay operations and as a result may allow invalid relationships between data products. Index-based approaches are also common
in digital image processing (Kauth and Thomas, 1976; Tucker, 1979; Crist
and Cicone, 1984), and the reliability of such transformations depends on an
adequate match between data calibration and applied mathematical manipulations. Measurements in a GIS may be categorical, ordinal, interval, or ratio in
nature (Harvey, 1969). The characteristics and valid mathematical operations
for these data types are summarized in Table 6.1. Those operations indicated
as valid in the initial entries of Table 6.1 will also be valid for the data types
listed subsequently. The accuracy of index-based methods is also dependent

on weighting schemes that accurately reflect the process under study. Whereas physical processes are often statistically parametrized, the creation of valid weighting schemes for representing social value systems, such as desirability, may be quite difficult. Such weighting schemes must be derived from unbiased, informed consensus and are generally difficult to translate into an interval or ratio measurement scale.

Information derived from spatial data products through statistically based analyses will be constrained by the assumptions of statistical techniques, which may in turn be confounded by the effects of spatial autocorrelation. Although the assumptions of various statistical methods are beyond the scope of this chapter, selected problems that are common with spatial data will be mentioned. Social and environmental data often violate the assumptions of multivariate normality required by classical statistics. Studies have demonstrated how inappropriate assumptions of multivariate normality in spectral data reduce the accuracy of automated land cover classification techniques (Maynard and Strahler, 1981; Skidmore and Turner, 1988). Curran and Hay (1986) demonstrate how measurement error in remotely sensed data may cause biased estimates in regression models for landscape parameters. This problem of error in regressors is generalizable to predictive relationships derived from map data. Multicollinearity – the existence of linear relationships between explanatory variables – is also common and presents a problem in regression modeling (Montgomery and Peck, 1982). In cases of multicollinearity, variance estimates for regression weights derived from ordinary least squares are inflated, resulting in potentially unstable values. As mentioned in Section 6.1, multicollinearity presents the danger of mistaking causation for correlation. Heteroskedasticity – the dependence of error variance on the magnitude of a measurement – frequently occurs in social and environmental data as well. Although there are robust techniques that may be useful in dealing with heteroskedasticity, this situation still provides problems in terms of efficient parameter estimation.

Finally, spatial autocorrelation, the tendency of proximate samples to have similar values, is practically universal in spatial data. This condition may violate the independence of samples required in classical statistics, resulting in underestimated sample variance and inflated confidence estimates. Techniques used to characterize spatial autocorrelation may include summary statistics such as Moran's "I" or graphic approaches such as semi-variogram or block variance analyses. The effects of spatial autocorrelation have been shown to reduce the accuracy of statistical land cover classifications when representative samples are not randomized (Craig, 1979; Campbell, 1981;

Labovitz, 1984). These effects have also been studied as they relate to local image variance (Woodcock and Strahler, 1987; Jupp, Strahler, and Woodcock, 1988, 1989), biophysical parametrization (McGwire et al., 1993), and error assessment (Congalton, 1988ab). Methods based on these findings should be developed to improve digital classifications, drive sampling methodologies, and deflate confidence estimates. The general lack of knowledge of methods for working with spatial data and a lack of integrated statistical tools within existing software packages are major impediments to error assessment in the analysis phase. The development of flexible statistical tools that take into account the particular difficulties of spatial datasets and the organization of these tools into a usable software environment may encourage adequate consideration of statistical assumptions in the development of higher order information products. Work in this area is currently being pursued with software such as the SPACESTAT package developed through the National Center for Geographic Information and Analysis (Anselin, 1992).

Though direct process modeling of social and environmental phenomena involves difficulties with simplification and suitable specification of boundary conditions and forcing functions, this approach may be more theoretically sound and generalizable than empirical methods based on statistical analyses. To explicitly model physical or social processes, GIS data are often exported to specialized, discipline-specific software. Results are then imported again for integrated assessment. Examples of such simulation environments include ties between hydrological data in a GIS and ground water models (Nystrom et al. 1986; Foresman, 1984) or links between crop classification strategies for imagery and econometric modeling (Schultink, 1982). The ability of process models to pass some indication of accuracy for derived products back to the GIS will vary. Simplifying assumptions, such as assuming independence of error between model components, may allow the propagation of error variance through a model to be estimated. Kerekes and Landgrebe (1991) provide such an example in their simulation of remote sensing systems. However, for complex nonlinear models or in cases where simplifying assumptions are not reasonable, the ability to estimate error propagation through a model may be limited to generic sensitivity analyses. In such cases, GIS and remotely sensed data may go beyond their role as a data source and might even be used to calibrate process model outputs. For example, Maas (1988) describes the use of spectral data to keep climatically based agricultural yield models on track with actual field conditions.

Process-oriented relationships between spectral reflectance and surface parameters are being studied in an effort to increase estimation accuracy for land surface parameters. Such efforts include development of directly invertible models of radiative transfer (Goel and Grier, 1986a, b, 1988; Sellars, 1985) and spatial variance (Li and Strahler, 1985; Franklin, 1988). One of the major difficulties with invertible modeling approaches is the requirement for large volumes of data (e.g., multiple look angles, ground surface characteristics, etc.). A possibly more tractable approach that is being tested estimates the physical composition of pixels through absorption features as measured by high-resolution imaging spectrometers. Examples include identification of surface mineral composition and plant canopy chemistry through distinctive spectral absorption features (Huete, 1984; Swanberg and Peterson, 1987; Kruse, Calvin, and Siznec, 1988). Inference of more abstract features in image data, such as land use, requires a complex understanding of natural and cultural systems. Expert systems and contextual classifiers have been tested to identify such features in image data (e.g., McKeown, Harvey, and McDermott, 1984); however, the complexity of the knowledge domain required to identify abstract features limits application of these methods to very specific tasks.

6.6. Assessment

As evidenced in the preceding sections, spatial data processing is an abstraction in which care must be taken to ensure that actual relationships in the system under study are accurately estimated. Therefore, it is critical that the validity of derived information products be tested to provide a reasonable estimation of confidence for use in decision making. Accuracy information is required in a decision-making process in order to understand the risk involved in relying on GIS-based information products. Such information may be of great importance in selecting between alternatives with respect to a particular risk-taking behavior. Part of the challenge of this accuracy assessment lies in direct quantification and visualization of product error (Beard, Buttenfield, and Clapham, 1991). Indirect assessment methods also play a valuable role in ensuring that high-quality information is produced by the spatial data processing flow. These indirect methods include conceptual and empirical models of the sources and propagation of error, as well as its impact on subsequent decision making.

The type of accuracy assessment required may depend on whether the results are relative or absolute measurements. In cases where simple information on distance or area is derived from a single data source, error such as a simple coordinate offset may not be significant. However, as described in the previous section, information derived from multiple spatial data sources will generally require enforcement of absolute positional accuracy. A similar dichotomy applies to thematic accuracy assessment when derived information products are either interval (relative accuracy) or ratio (absolute accuracy) in nature. Some authors suggest dividing accuracy assessment in GIS operations between attribute and locational components (Vitec, Walsh, and Gregory, 1984; Walsh, Lightfoot, and Butler, 1987). Such a division may be naive for GIS and remote sensing representations of continuous variation in fields, since the spatial objects that populate the database are to a large degree artifacts of the process of representation.

There is a large body of literature on direct thematic accuracy assessment in remote sensing (Hord and Brooner, 1976; Card, 1982; Aronoff, 1982a, b; Congalton, Olderwald, and Mead, 1983; Rosenfield and Fitzpatrick-Lins, 1986), and an extensive collection of these articles has been compiled by Fenstermaker (1994). These efforts use contingency matrices to compare database contents with samples derived from ground survey or some other information source in which there is a high degree of confidence. These matrices provide detailed information on the types and magnitudes of error found in original data or derived information products. In remote sensing classifications, the matrix typically relates the class assigned to a pixel in the database with the class determined for the same pixel by ground survey (per-point assessment). In many GIS applications, which use the polygon data model, the error matrix may compare the class assigned to an entire polygon with the class assessed by visiting the polygon in the field. This per-polygon assessment clearly omits within-polygon variability from the definition of accuracy.

The simplest statistic derived from the contingency matrix is the percent correctly classified (PCC) or the percent of cases falling on the diagonal of the matrix. The matrix may also be examined using row or column aggregates to test the accuracy of the map product with respect to estimated errors of production and subsequent use (Aronoff, 1982a; Story and Congalton, 1986). Row and column statistics may also provide user and producer accuracies for individual classes, but they lack the sensitivity to describe cases where accuracy is strongly dependent on confusion between specific class pairs. These dependencies will be represented by off-diagonal entries in a contingency matrix.

Because some points will be classified correctly by chance even in a random assignment of classes, PCC is often rescaled to discount this effect, yielding the kappa statistic (Congalton et al., 1983; Rosenfield and Fitzpatrick-Linz, 1986). The kappa statistic is sensitive to the off-diagonal entries of a contingency matrix. In addition, the distribution for the kappa statistic is asymptotically normal; thus, the significance of differences between alternate map products may be tested (Congalton et al., 1983). At present, no analytic connection has been made between kappa and conventional measures of positional map accuracy, such as epsilon. Despite the utility of both the PCC and kappa statistics, these measures reduce the dimensionality of error characterization to a single metric and may never adequately describe products with variable class accuracy. Basically, any reduction from the full contingency matrix to a smaller set of representative statistics reduces information content. Thus, presentation of the full contingency matrix along with thematic data products may be required for proper assessment of product accuracy or suitability.

Methods based on error matrices assess representations for fields of categorical variables, such as land cover class or soil class. Although this accounts for much of the information in GIS databases, it is also important to assess fields measured on continuous scales, such as spectral response or topographic elevation. This class of error estimation has been addressed in both GIS and remote sensing literatures. For example, McGwire and Estes (1987) compare the error assessment capabilities of moving-average and kriging interpolation methods. Using cross validation, a single error statistic may be generated from moving-average interpolations. In contrast, kriging yields a field of error estimates, providing a better understanding of the positional dependence of uncertainty. Figure 6.3, taken from McGwire and Estes, presents an example of the accuracy assessment made possible by kriging. Whereas cross

Figure 6.3. Error assessment using the kriging technique

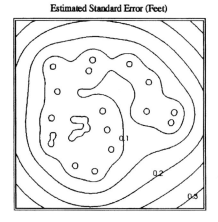

| Interpolated Ground Water Elevations (Feet) | Estimated Standard Error (Feet) |

validation of a moving-average interpolation provided a product with a single standard error estimate of 0.81 ft, the kriging method reduces uncertainty and shows how error would be expected to vary throughout the site. Atkinson (1991) demonstrates the use of a similar technique for determining an accurate estimate of the unbiased mean of a continuous variable for samples within a pixel.

In addition to assessing the accuracy of attribute measurement within a field, it may also be important for the decision maker to understand the positional accuracy of features. Uncertainty in the placement of a point can be characterized as a two-dimensional probability density function centered on the observed location of the point. It is commonly assumed that errors in the x and y directions are uncorrelated and normally distributed with the same variance, so that the distribution is circular normal. A common statistic based on this distribution is the circular map accuracy standard (CMAS), which is defined as the radius of a circle centered on the observed point and containing the true location with 90% probability, or more generally, the 90th percentile of the distribution. CMAS is commonly used as the basis of map accuracy standards, including the National Map Accuracy Standard of 1947. Other statistics, such as the root mean square error (RMSE) and the mean square positional error (MSPE), are also used to describe the probabilistic position of points. The majority of image processing and GIS software packages derive RMSE as a diagnostic of the geometric transformations used to coregister data. However, this diagnostic should not be confused with an independent accuracy assessment because test points are not independent of the transformation parameters. The resulting RMSE value is therefore likely to provide only a best case estimate of positional error. In contrast to point data, no satisfactory models of positional uncertainty currently exist for lines. Uncertainty in the position of linear features must be handled differently from points because adjacent positions along a line are not likely to be independent (Keefer et al., 1988), because the direction in which displacement occurs becomes ambiguous with respect to the x or y coordinate dimensions, and because it is possible to match observed and true locations only in limited circumstances. As mentioned in Section 6.3, uncertainty in boundary location has been described, but not modeled, using the epsilon band approach (Perkal, 1956). Blakemore (1984) demonstrates such use of the epsilon band to return uncertain responses to queries about the containment of a point within a polygon.

In order to track error accumulation effectively, methods are required to assess the generation of error associated with specific processes and to keep an accounting of the spatial, temporal, and attribute characteristics of this accumulating error. Methods for providing a transcript of data processing histories exist in many commercial remote sensing and GIS packages. However, this capability has generally been unsophisticated and has not allowed for specific inclusion of error characteristics. An integrated solution to tracking the data processing flow, called lineage tracing, is described by Lanter (1989). This approach uses a LISP language shell in which the Arc/Info GIS package (produced by Environmental Systems Research Institute) is run. The described algorithm allows automated backwards and forwards reconstruction of intermediate data products between data inputs and information outputs. Ongoing research is focusing on the application of this technique to modeling error accumulation in GIS information products (Lanter and Veregin, 1990).

Clear communication of spatial data using graphic and text products is also critical for accurate representation of information to decision makers. Much of the work in this area has already been addressed by the long tradition of manual cartography. However, whereas traditional cartography has focused on the legibility and perception of geographic information, recent recommendations suggest that visual representations of uncertainty associated with geographic data be provided as well (Beard et al., 1991; Lunetta et al., 1991). In going beyond these visually oriented approaches, the relationship between map accuracy and specific information requirements of the environmental and policy domains must be evaluated. Accuracy of information products must be evaluated in relationship to the risks of a management decision, whether regarding agricultural production or deforestation policy. This evaluation might be explored in the context of operations research methodologies for multiobjective decision making (e.g., Hobbs and Voelker, 1978). Figure 6.4 plots the attractiveness of possible alternatives with respect to two different objectives. Examples of these objectives might be reducing environmental impact versus cost of operation. In this graph, a curve of noninferior alternatives may be identified that represents the trade-off between objectives. Uncertainty may vary between the information products used to quantify these objectives. As a result, the position of points relative to the noninferior curve is probabilistic and identification of noninferior sites may be less obvious. At present, methods for incorporating uncertainty in multiobjective decision making are not well developed (Solomon and Haynes, 1984). However, such techniques may benefit from the formalization of a comprehensive model for accuracy assessment.

6.7. Conclusion

Our understanding of accuracy issues in spatial data processing has yet to
be fully described within an accepted, coherent framework. Several sources
provide intensive investigations of error sources and modeling. Kerekis and
Landgrebe (1991) have simulated remote sensing systems to the point of
predicting the effects of spatial autocorrelation on supervised land cover classifications. Veregin (1989) provides an excellent organization and review of
error assessment and modeling techniques. However, the goal of a coherent
system that integrates efforts such as these has proven elusive. Such a conceptual framework would allow better understanding of an information product's
"fitness for use" (Chrisman, 1991) in a given application. At present, the
level of understanding of accuracy issues in the research community is well
in advance of the corresponding level of understanding in current practice.
Unfortunately, significant improvement in the accuracy of spatial data or in
the ways uncertainty is dealt with in practice will only occur at substantial

Figure 6.4. Plot showing variable confidence in quantifying objectives

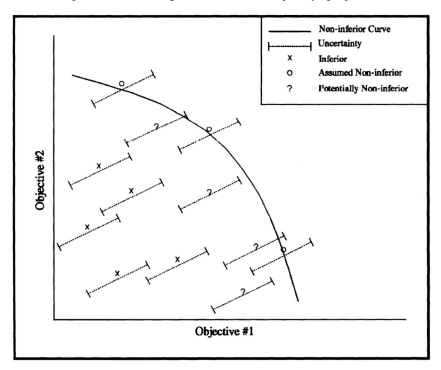

cost. Nevertheless, computational capabilities have been increasing at a dramatic rate, and the additional processing required by methods for modeling and visualizing uncertainty is increasingly manageable. Further research may also develop more efficient methods for characterizing accuracy and resolving uncertainties.

Several areas require further investigation if accuracy is to be improved or error characteristics better understood. In remote sensing, technical developments such as advanced geometric rectification capabilities and feature extraction methods will increase the quality of data that may be brought into GIS-based analysis. Accuracy characteristics will vary between data products, and metadata management capabilities must be developed to make these characteristics accessible to users and error tracking processes. Methods for understanding the effects of spatial data processing on product accuracy and managing these effects in ongoing analyses are required. As an example, a system might alter information on the spatial precision of a secondary data product that is generated by a neighborhood operation. The challenge of statistically characterizing spatial processes also stands, both in developing accurate predictive relationships from map-based variables and in describing the interdependent roles of thematic and positional error. Methods for using accuracy information in spatially oriented decision making must also be better developed in order to understand specific risks associated with using information products.

An alternate approach for developing an accuracy model might focus on decision-making requirements rather than the specific processes that create error. This approach would focus on specifying the minimum set of spatial and aspatial descriptors for information product accuracy that would be required in the decision-making process. The resulting standardization of information required from various GIS software packages might then highlight existing deficiencies and provide impetus for further development of critical components for internal representation and manipulation. This developmental approach emulates that of the relational database model, which in its early stages was specified by its functional appearance to the user rather than its internal structure and manipulation language (Date, 1986). To date, few relational databases are totally compliant with the full conceptual development of the relational model. However, specification of consistent, yet flexible, user interaction promoted wide acceptance and spurred interest in further development.

7

Integration of remote sensing and GIS technologies for planning

T.W. Foresman and T.L. Millette

7.1. Introduction: Planners and spatial data evolution

Planning is fundamentally a problem-solving profession that deals with the factors that influence the quality of life at scales ranging from neighborhoods to nations. Planners use a wide variety of data and analysis techniques to support the policy decisions related to an ever changing landscape. The process of what is generally termed planning is applied to a wide variety of activities that range from discrete local activities, such as monitoring building permit compliance by a homeowner remodeling a garage, to development of a sweeping ten-year regional master plan for a rapidly developing rural county or perhaps an entire state. The enormous differences in the scales at which planning activities take place roots the process in the understanding of geographic entities and therefore in some form of spatial data analysis. Through comprehensive planning to link the human and physical environments and by applying methods of data analysis, planners attempt to explore solutions for community problems.

If you consider the basic nature of what planners do on a day to day basis, the environment in which they apply their craft, and the tools with which they work, the variety is broader than many traditional disciplines. County planning agencies in rapidly urbanizing areas such as Montgomery County, Maryland employ several hundred professionally trained planners with technical specialties that include transportation, solid waste, urban hydrology, housing, land use, urban forestry, economic development, and many more. Rural planning agencies such as the Bennington County Regional Commission in southern Vermont have four full-time planners to cover the same breadth of activity. However, common to both organizations is the need to understand the geography around them and to have a firm grasp of the landscape in which they live, be it forest or concrete. Further, they must understand the community's role in the landscape as an invaluable starting point from which to develop strategies to protect and improve the quality of life for all residents.

Planners continue to refine their geographic understanding of landscape and population through a formal process of data collection and data analysis. The American Planning Association (1981) states "Urban and regional planning is a systematic, creative approach and method used to address and resolve social, physical, and economic problems of neighborhoods, cities, suburbs, and metropolitan regions."

Urban planners have relied primarily on census information supplemented with local government statistics. Rural and regional planners, however, have relied more heavily on aerial photography with census data and field surveys to assess less populated areas. Difficulties sometimes result due to errors in the collection and processing of geographic data. These inaccuracies together with data handling errors can cloud the true nature of the social and/or environmental milieu. Comprehension of the milieu is the starting point from which it becomes possible to develop analytical strategies addressing the myriad of urban and regional problems. It is the nature and magnitude of these problems that motivates planners in their search for technological assistance.

7.2. Early technology and planners

Historically, planners have been active consumers of aerial photography and remotely sensed imagery from the inception of these technologies. Planners' dependence upon maps for the geographic context of their analyses forged an early natural alliance with the photogrammetric and remote sensing communities. For example, French urban planners in the late 1800s found topographic maps made from photography to be as accurate as their "traditional" field survey maps. The basic principles on making maps by using photographs taken from balloons were patented by C.B. Adams of the U.S. Army in 1893 (Whitmore, 1952). Planners continued to depend upon and utilize advances in photogrammetry and mapping technology throughout the twentieth century. By the late 1970s, the remote sensing community experienced significant improvements in spatial, spectral, radiometric, and temporal resolution of satellite imagery. Access to this remotely sensed imagery significantly enhanced planners' opportunities for mapping urban spatial structure and tracking urban sprawl (Holz, Huff, and Mayfield, 1969; Jensen et al., 1983), in addition to a host of rural land use and natural resource investigations used to facilitate regional planning.

The advent of the electronic revolution has brought many important advances to the planners' tool box. Digital computers have become important tools for the compilation, analysis, and production of an extensive array of planning products. Various analytical models, designs, maps, and reports produced by planners depend entirely on the use and availability of computers. Computing technology growth has provided planners access to a broader range of analyses and has also provided access to vast amounts of spatial referenced digital data. As both map makers and map users, planners are reaping the benefits of the explosive growth of spatially referenced digital data from both governmental and commercial sources. Spatial data from global positioning satellite systems and digital cartography provide a wealth of resources. Planners' use of geographic information systems is being fueled by the combination of increasingly affordable computer technology and digital data availability.

Planners have embraced GIS technology and methodologies from their earliest development. The 1838 *Atlas to Accompany the Second Report of the Irish Railway Commissioners* (Parent and Church, 1988) exemplifies planners' use of base maps with mapping overlays. Streich (1986) credits Herman Hollerith for providing the first automated demographic database by using punch cards for the 1890 United States Census. This event began a practice for automating planning information that continues to today's decennial census data. Canadians are recognized for the development during the mid-1960s of the Canadian GIS (CGIS) for national and regional agricultural land inventory and planning (Peuquet, 1977). By 1968, just four years after the CGIS was implemented, urban and regional planning use of automated information management systems was marked by the installation of thirty-five systems in local governments throughout the United States (Systems Development Corp., 1968). Today, thousands of GIS installations are used by planners, as demonstrated by GIS hardware and software sales and the increase in technical and policy-related conferences.

7.3. Planning and GIS

The process of integrating map-based spatial data for landscape planning was first articulated by McHarg (1969) in his book *Design with Nature*. Maps of slope, surface drainage, soil drainage, bedrock, soil foundation, and erosion susceptibility were combined in analog overlays to create a composite of "physiographic obstructions" for route selection of the Richmond Parkway in New York. McHarg used this composite of physiographic obstructions, together with a composite of social values developed in the same manner, with map layers that included land value, historical value, scenic value, recreational value, residential value, water value, forest value, and wildlife value. This "ecological method" of route selection using data reflecting social, resource, and aesthetic values (McHarg, 1969) was an important moment both in planning and in the activity that would later be collected under the moniker of geographic information systems. In planning, McHarg's book was a harbinger of an important new paradigm that put a broader collection of social ethics ahead of what was all too often a narrowly defined set of engineering and cost considerations for development. In GIS, it presented Boolean operations as a useful method of integrating multilayered map-based information. More than two decades has passed since *Design with Nature* was first published, and in that time an entire industry of GIS hardware, software, consulting, education, data capture, and database development has exploded around us. Boolean operations still remain the most frequently used method of multilayer spatial data modeling for planning.

7.4. Planners, activities, and data

Although the realms and activities of planners are both broad and numerous, it is useful to attempt to generalize them into a core set for the purposes of this discussion. If we consider those activities that are common to most planning organizations, these are likely to include issues pertinent to land use, land cover, infrastructure, zoning, transportation, master plans, census demographics, environmental protection, and regulation. A common characteristic of these basic planning activities is that they all require the use of spatial data, or more specifically, data that are linked to discrete locations. In general, what distinguishes planning from other data intensive activities is that virtually all planning data is related to geography and location (Cooke, 1980; APA, 1981).

Specific tasks that are required on a daily basis in most planning agencies include:

• Inventory and analysis of many types of spatial information, such as changing patterns of land cover and land use to support master plan development.
• Administration of zoning bylaws, which requires determining the location of a parcel in relationship to established zoning districts.

- Site plan reviews, which examine site conditions such as soils, slopes, ground water, utilities, setbacks, right of ways, and variances to ensure adherence to code.
- Subdivision reviews that identify the potential impacts of residential development on community facilities and services.
- Permit tracking to identify the location of building permits issued to determine if development is conforming to master plans.
- Property assessment based on listing requirements such as parcel size, location, address, frontage, building size, condition, and use.
- Facility siting, which involves identifying site conditions and proximity to particular land uses, utilities, and transportation routes.
- Infrastructure maintenance to set up a historical database and inventory of installations and provide for inspection and maintenance of roads, bridges, culverts, and sewer, water, and energy utilities.
- Event reporting of crime, fire, and accidents to be used for resource allocations of personnel and equipment with a strong spatial rationale.
- Dispatching of emergency vehicles based on route optimization, given travel distance, road conditions, traffic patterns, and speed limits.
- Vehicle routing for school buses, snowplows, wide loads, and hazardous materials to balance expedience against public safety.
- Disaster preparedness to preposition strategic materials, prepare evacuation routes, and identify high-priority facilities including hospitals, schools, and water supplies for emergencies.
- Traffic analysis to identify origins and destinations of traffic at peak hours to minimize congestion and gridlock.
- Legal notification for public hearings that require notification of abutters and land owners within a specific radius of a particular parcel.
- Acquisition and disposal of property for the purposes of condemnation, foreclosure, and right-of-way acquisition. This requires information on roads, buildings, land use, zoning, ownership, and adjacent parcels.
- Tax listing, revaluation, and abatement programs for management of local land and tax bases.
- Redistricting and rezoning for schools and political representation (Dangermond and Freedman, 1986; Budd, 1990; Dueker and DeLacy, 1990; Millette, 1990).

Imagine trying to accomplish any of these tasks without the ability to integrate spatial information from a variety of sources such as a community's parcel map, road database, census data, master plan, zoning map, land use inventory, or even up-to-date aerial photography or remotely sensed imagery.

How planners go about the process of planning has been aptly described by Cowen and Shirley (1991, p. 297) as confronting "a myriad of ad hoc decisions which require accurate and current spatial data." These decisions are framed by the overall planning goals and objectives and constrained by practical limitations on time and resources under which all planners toil. Specific planning goals such as growth management, agricultural land preservation, economic development, or urban restoration provide planning organizations with a perspective and an agenda from which to operate. From these goals are fashioned a set of concrete analysis objectives that determine the modes of assessment and the models and data required to support them. GIS has become an invaluable planning tool because it, at this time, is the most productive environment for processing spatial data. This increase in productivity allows planners to conduct more robust assessments than would be otherwise possible given limitations of time and resources.

The use of spatial data to support the wide spectrum of planning activity can be grossly categorized into one of two classes of operations. The first is the use of data representations of discrete features or objects such as a road, bridge, or building. The second is the use of data representations of continuous surfaces such as soil types or land cover classes. Planning activities that require working with discrete features in a GIS environment generally represent them with the vector (point) data model, whereas those requiring the representation and processing of continuous surfaces generally make use of the raster (cellular) data model. Although data can be moved between these environments by automated raster-to-vector conversions (Waters, 1989; Sakashita and Tanaka, 1989), analysis is more difficult between rasters and vectors. Therefore, planners capture data and process it in the form that is most efficient and convert between rasters and vectors when appropriate. For example, it is becoming routine for planners to use raster-based satellite multispectral data and image processing to develop and update land cover and land use databases and subsequently move these raster classifications into the vector environment by conversion so that they can be used with a variety of other GIS data (Gernazian and Sperry, 1989; Stow et al., 1990; Millette and Sickley, 1992). However, the key to successful implementation of spatial data processing and GIS in planning activities is to identify the planning goals and objectives, create assessment programs that will meet these objectives, identify models and data required to conduct these assessments, and convert the data into formats that are compatible with the GIS being employed in the particular agency. A typical set of basic planning GIS layers would include parcels, roads and transportation corridors, surface waters, land use and land cover, zoning, building structures, soils, and topography (Figure 7.1).

7.5. Basic components of integrated spatial information systems for planning activities

Spatial databases used by planners require constant updating to record changes in a wide variety of attributes, including but not limited to changes in owner-ship, population, construction, and land cover. To effectively address manage-ment of planning databases, principles of information systems management should be used. The organization of data handling and analysis activities using principles of database management can be accomplished with what are termed "systems management" steps. These essential information management steps as applied to GIS and remote sensing technology are depicted in Figure 7.2.

7.5.1. Collection, input, and correction

A host of remote sensing information and other data useful for planning has been demonstrated by a number of researchers (Holz et al, 1969; Jensen et al,

Figure 7.1. Conceptual GIS planning layers

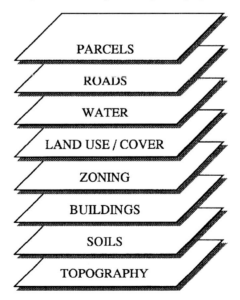

PARCELS

ROADS

WATER

LAND USE / COVER

ZONING

BUILDINGS

SOILS

TOPOGRAPHY

Figure 7.2. Integrated planning functions using GIS and remote sensing technology

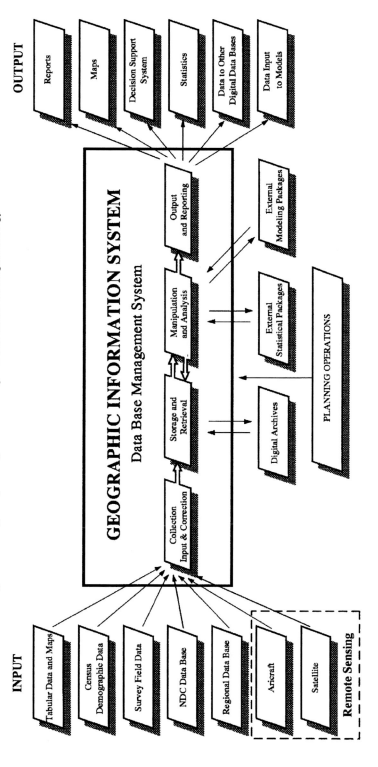

1983). Automated handling of these data requires acquisition from multiple sources together with a series of preprocessing steps to make them accessible for analysis. Systems that use combinations of ASCII text files and vector and raster graphic formats require sophisticated software and hardware and highly trained human operators. The complexities of data handling and conversion limits many planning organization's access to the full range of data currently available. Innovations by commercial data brokers are beginning to address these limitations by offering enhanced preprocessed data products. For example, satellite data is currently being repackaged by *SPOT* and *EOSAT* into affordable data packages of state, county, and quadrangle scenes for ease of use by state, regional, and municipal agencies. Other commercial vendors offer geometric and terrain corrections for satellite scenes. A variety of data brokers offer enhancements for regional subsets of census data. Spatial data transfer standards (SDTS) as promoted by the federal government should improve access to these various forms of preprocessed data. Additionally, commercial standards are evolving that provide some assistance for the integration of multiple spatial data formats.

Data vendors that specialize in georeferenced data sets are becoming aware of the market needs for higher quality data. Data accuracy labeling has been a growing issue as both multiple use of shared databases and data liability awareness increase (Foresman, Kelley, and Shalit, 1990). Development of automated quality assurance (QA) procedures during the data collection phase will be needed to further control data integrity.

7.5.2. *Storage and retrieval*

Data storage and retrieval efficiency play a vital role in the handling of spatial data. Over the past decade much GIS data has been in the form of vectorized point, line, and polygon graphic files linked to socioeconomic and demographic tabular data. The data volumes associated with vector databases have been extremely large as a result of the explicit nature of topologic data structure. Planners wrestling with complex landscape analyses, however, have found applications that use raster images of land cover and terrain as backdrops with vector overlays of administrative districts or parcels to be useful. Raster-based GIS systems are increasing in popularity due to greater availability of digital orthophotography and satellite imagery, although these data are storage intensive due to the large arrays of pixels associated with landscape imagery. Although raster software packages offer planners increased functionality to integrate remotely sensed data with the political and administrative data, they are constrained by digital storage media for access and retrieval. Advances in digital storage media appropriate for large raster and vector datasets include such options as digital audio tapes (DAT), magnetic disk, Bernoulli boxes, CD-ROM, write-once-read-many (WORM) drives and optical disks. These advances in storage media indicate that storage and retrieval should keep pace with the rising demands of GIS databases.

7.5.3. *Manipulation and analysis*

Planners frequently require the ability to process geographic data and models and bring detailed analyses into the policy arena. Spatial information developed during the planning process may be an end in and of itself or part of a larger process leading to further projections and analyses.

Generally, planning analyses build upon existing knowledge and databases, project trends and options, and continue to refine the understanding of physical and human characteristics of a community.

Geographic data management provides a tool for conducting analyses and appraisals of proposed actions. Models of the urban structure or rural setting can be based on measurements taken from remotely sensed data. In a GIS context, these data allow visual appreciation of the geographic setting and review by administrative units of thematic components such as:

- areal extent and population,
- urban size and distances between urban centers,
- multijurisdictional divisions and overlaps,
- transportation networks,
- commercial and industrial nodes, and
- physical and environmental composition.

Many planning organizations presently make use of remote sensing and GIS technologies at this basic level of "display analysis." More sophisticated planning organizations move beyond display analysis into more robust analytical assessments.

Advanced analysis begins with accurate inventories of current conditions. Application of analytical techniques, including regional forecasting, spatial filtering, and map generalization, builds upon this basic inventory. Trend analyses can be applied to subdivide original planning variables into additional interpretive subcomponents. Aggregations based on administrative units are commonly used to plot data at centroids that are subsequently used to generate isopleth surfaces. Various interpolation methods, such as kriging, can be applied to represent mapping surfaces from discrete and often undersampled data sets. Planning analyses commonly require linking statistical routines to spatial databases, through importing and exporting files, or the use of built-in statistical routines contained in commercial GIS software. The inherent limitations of statistics components in most GIS software is an important but often overlooked consideration. As a result, planners frequently use dedicated statistics software such as SPSS or SAS.

Linking GIS to analytical planning models represents a more complex level of the manipulation and analysis. Terrain modeling, for example, is an important component for many planning studies. Use of polyhedral techniques, which approximate a surface using a system of triangular elements, are used for calculating slope, aspect, and runoff values. These terrain values are important inputs for many numerical models. The grid structure provides a linkage format to integrate different types of numerical and spatial data. Processed satellite imagery combined with administrative units and other areal data can be incorporated together for modeling by utilizing the grid data matrix. Experience with integrated GIS and remote sensing has significantly impacted the way planners design their spatial databases and use modeling software. A typical planning organization will encounter the need for quantitative land use/land cover change detection; social, economic, and environmental evaluations of population; spatial autocorrelation for enterprise development zones or renewal zones; pollution assessments for air sampling; water quality and hydrologic runoff modeling; recreational and open space siting; wetland reviews and permitting; and landfill siting. Whether in a four-person or two-hundred-person agency, the need to conduct these analytical processes remains essentially identical. Therefore, the structure of the spatial database and the ability to work with the appropriate external modeling packages becomes a major driving force for research and development by vendors, academia, and planning organizations alike.

The list of common external planning models being linked to commercial GIS and remote sensing software is growing. Examples include TRANSPLAN for transportation analysis, MODFLOW and SURFACE II for hydrologic runoff modeling, and CAMEO for emergency air plume assessment. As these and other modeling packages become linked to commercial geoprocessing software, more standards for operating procedures and reporting formats can be expected.

7.5.4. Report generation – communication

Communication is a critically important planning function. Planners typically participate in a wide range of policy and public forums that can be well served by the graphic and reporting products of GIS. For example, forecasting electrical load demands for a region may require planners to reformat information on discrete point or areal geographic objects; tabulate daily, annual, or decade per capita consumption rates; assess environmental impact projections to citizen groups; and compile these findings in a report to the public utility commission. GIS outputs have enabled many planners to keep pace with their ever-expanding reporting demands.

Basic output requirements for planning organizations include adequate computer monitor display, printing, and plotting (see Chapter 4). Mapping output sizes should include a range from A- to E-size plots. A series of basic planning maps should include a variety of topographic and thematic GIS layers. Plots of statistical summaries, graphs and charts, and choropleths depicting various geographic datasets are also requisite to communicating planning results.

Visualization of the landscape or urban structure may include use of three-dimensional contouring, with surface modeling overlays. Satellite imagery draped on topographic elevation models provides enhanced appreciation of visibility issues, viewsheds, and the general spatial context. Interpolations and other surface generalizations can be mixed with the raster satellite images for a better understanding of complex spatial patterns that would otherwise be indecipherable to planning audiences and decision makers.

It is important to consider the basic character of GIS databases integrated with grid-based satellite imagery. These spatial data structures represent both the enabling and limiting factors in transforming planning analyses into cartographic products. These output products are important to the intelligible display of complex planning results.

7.6. Current issues in technology usage

Remote sensing is increasingly being considered as one element of an integrated GIS environment rather than simply as an important source of thematic multispectral data to be moved from the remote sensing domain into the GIS domain (Jackson and Mason, 1986; Parker, 1988; Ehlers, Edwards, and Bedard, 1989; Zhou, 1989; Star and Estes, 1990). For the application of remote sensing and GIS to planning, there are a host of considerations necessary to ensure that multispectral data can be successfully converted to information useful for planning. Included in these considerations are issues of temporal fidelity, scale and resolution, quantitative versus qualitative model integration and linkage, operational linkages, costs, training, competency, computational overhead, and standards.

7.6.1. Temporal considerations

The ability to reliably detect and identify specific materials representative of information categories from an aerial or orbital platform is significantly affected by the temporal characteristics of the data captured by a sensor or film type. For example, the phenology of a particular crop type or vegetation species plays an important role in selecting the optimum time for capturing remotely sensed data. Most planners know that early spring is the ideal time to distinguish land cover due to minimal canopy effect for most of northern America. Since many planning organizations lack the resources to use multidate, multispectral data for general land cover database development, it is necessary to identify the best time to capture data given the geographic setting and the features or land cover materials of interest. For example, automated extraction of roads from *SPOT* satellite data for 1:50,000 scale map updating in Ontario used spring imagery in part to minimize canopy effects (Maillard and Cavayas, 1989). For the purposes of modeling within scene variability, temporal relationships between different crop types has proven successful in improving accuracy when used in a knowledge-based approach to image classification (Middelkoop and Janssen, 1991).

Temporal characteristics play a role in detecting changes in surfaces over time together with spatial, spectral, and radiometric properties of sensors (Townshend and Justice, 1988). However, comparison of images from the same sensor captured at different times presents the analyst/user with a variety of problems associated with sensor performance and atmospheric conditions. The use of data from multiple sensor systems for change detection has the additional complications of differences in pixel size, spectral bandwidths, and spectral response properties (Duggin, 1985). For most types of municipal and regional planning applications the use of multitemporal satellite imagery is fairly limited except when extremely detailed land cover classifications are required. Although temporal considerations for remote sensing can be complex, from a planning perspective they can be distilled to a simple and practical rule of thumb: "timing is everything." Give a planner the choice of data with higher quality (e.g., spatial, spectral, or radiometric characteristics) or with better timing (diurnal, seasonal, year) and they will almost always choose the temporally stronger data. Studies involving resource managers and planners have shown that, given a choice, currency will be taken over accuracy in a variety of situations (NRC, 1990).

7.6.2. Scale and resolution considerations

Scale and resolution of photography and remotely sensed imagery have a pervasive influence on the types of information that a given user can extract. Scale is the ratio between the size of objects in the environment and their resultant representation on photographs, maps, or models. Scales of aerial photographs used for local planning purposes typically range from 1:200 for detailed engineering applications to 1:10,000 for more general resource inventory uses. Regional applications make use of photo scales up to 1:60,000. Many municipal planning organizations have access to orthophotography commonly used by the engineering or public works departments. Orthorectification removes topographic and geometric distortions from photos and registers the photographs to a coordinate grid such as the local State Plane or Universal Transverse Mercator (UTM) systems. Recent developments in orthophotography, scanning, and geoprocessing has led to increased availability of digital orthophotos (Dall, 1990; Shanks and Wang, 1992), which we believe will become more economical and popular in the next few years. Today, large-scale high-resolution aerial photography is generally converted to planning GIS layers and databases by manual interpretation and digitizing. For example, thematic information such as road networks, utility distribution networks, and housing stocks are typically photointerpreted and digitized by hand. When projects are large enough to require more than several person-months worth of digitizing, automated scan-digitizing generally becomes more cost effective.

When discussing digital satellite imagery, resolution rather than scale is the appropriate consideration. The term resolution is actually a multidimensional concept that includes spatial, spectral, radiometric, and temporal considerations. For a complete discussion of resolution, see Simonett et al. (1983). For the purposes of this discussion, it is the spatial characteristics of satellite data that are particularly germane.

The term spatial resolution when applied to sensors is problematic since the term itself can be defined in a number of different ways. For example, what most often is termed the pixel (picture element) size (the ground covered by a discrete reflectance value) is actually the instantaneous field of view, which is a product of the sensor's optical system, the size of the detector element, and the altitude of the instrument at the moment of imaging. The effective resolution element (ERE) is the smallest object that can be reliably detected against a spectrally contrasting background, taking into account atmospheric conditions, sensor and object geometry, spectral contrast within the scene, and any enhancements to the imagery after it was recorded (Billingsley et al., 1983; Forshaw et al., 1983; Duggin, 1985; Strahler, Woodcock, and Smith, 1986).

For local planning purposes, applications of satellite data are generally restricted to sensors with relatively fine spatial resolutions, such as *SPOT* Panchromatic, *SPOT* Multispectral, and the *Landsat* TM. These sensors have nominal spatial resolutions of 10 m, 20 m, and 30 m, respectively. In an attempt to combine the spatial characteristics of *SPOT* Panchromatic with the spectral characteristics of *SPOT* Multispectral or *Landsat* TM, a number of techniques have been developed to combine and enhance multiresolution imagery; these include the hue-intensity-saturation (HIS) (Welch and Ehlers, 1987; Thormodsgard and Feuquay, 1987), principle components analysis (PCA) (Chavez and Kwarteng, 1989), and high-pass filtering (HPF) (Chavez and Bowell, 1988). The HPF approach creates the least distortion to the spectral data (Chavez, Sides, and Anderson, 1991). Spatial details in *SPOT* and TM data have been adequate to allow a wide range of planning applications, including road extraction (Maillard and Cavayas, 1989; Wang 1992), urban modeling (Meaille and Wald, 1990), regional growth analysis and local planning (Ehlers et al., 1990), rural land use analysis (Nellis, Lulla, and Jensen, 1990; Millette, 1992), rural-urban fringe analysis (Gastellu-Etchegorry, 1990; Treitz, Howarth, and Gong, 1992), urban impervious surface analysis (Dolan, Martin, and Warnick, 1984; Plunk, Morgan, and Newland, 1990), and urban land use database updating (Stow et al., 1990), as well as many others.

7.6.3. Data, model, and linkage issues

A fundamental reality of most planning assessments is that time and money are always limited. Therefore, the productivity of any tool under consideration in a planning organization is of paramount importance. Remote sensing has made a significant impact on productivity in a number of areas but most notably in land cover and land use GIS database development. Before the introduction of the *Landsat* series of multispectral satellites in 1972, most land use and land cover inventories were done from large-scale aerial photographs that were both time consuming and expensive to acquire and to interpret. An alternative approach termed "windshield surveys" required planners to drive around the community and map from visual inspection. Because the cost of acquiring and interpreting large-scale aerial photography is substantial, it was and is often beyond the means of many communities to conduct updates on a regular basis, if at all. Windshield surveys work fine in small areas with homogeneous land cover, little topographic expression, and road accessibility; however, for many communities this combination of characteristics simply does not exist. Frequently, planners are forced to use out-of-date photographs together with windshield surveys to try to develop contemporary inventories, often with mixed results.

The introduction of satellite data and image processing presented the planning community with the opportunity to efficiently and relatively inexpensively update land use and land cover databases such that they were systematically up to date and internally consistent. Remote sensing technology presented the opportunity for many communities to create land use and land cover inventories that could be used quantitatively as well as qualitatively. Changes in land inventories can be identified in terms of location and character with greater efficiency since the data are captured, enhanced, classified, and tabulated completely within a digital environment. The integration of remote sensing with GIS presents a wide range of opportunities for planning-related spatial modeling that until recently was prohibitively expensive due to the high cost of database development.

In addition to information directly extracted from digital satellite data, a plethora of new digital data sources are being made available to the planning community. Although not all these data are available throughout the United States at large scales, they will be in the future and hold important promise for urban and regional planners. Digital elevation models are useful for a wide variety of applications, including vertical zoning, soil, and vegetation mapping, and are particularly useful for removing differential illumination effects in satellite data caused by topography (Civco, 1989). The U.S. Census Bureau's Topological Integrated Geographic Encoding and Referencing (TIGER) System provides 1:100,000 scale digital street maps for the entire United States, with address ranges for streets in the 345 largest urban areas. The USGS Digital Line Graph (DLG) database (Allder et al., 1984) are digital representations of the 7.5-minute quadrangle series subdivided into thematic layers such as political and administrative boundaries, transportation networks, and hydrography. These digital cartographic databases provide valuable computer-compatible maps that allow planners to spend time analyzing spatial data rather than capturing it. Most notably, TIGER files allow planners to map and analyze socioeconomic data from the Census Bureau's decennial census for numerous types of demographic analysis, including population, housing, health care, crime, and many others. DLG digital maps provide an inexpensive option for communities to develop a basic collection of planning layers when used as a geographic base within a GIS environment. Because both TIGER and DLG are registered to map coordinate systems, they can be integrated

with aerial and orbital remotely sensed data and offer a potentially low-cost
path to operational GIS implementation.

The types of ad hoc evaluations and decisions made by planners generally
require a large variety of data and information. These data and information are
typically distributed throughout numerous governmental agencies and generally involve a wide range of scales, resolutions, tabulation units, coordinate
systems, and media formats. Many planners in the process of developing
a community's comprehensive plan have found themselves doing the "data
drill," having to obtain information from the public works department for
aerial photos, the highway department for road maps, the town clerk's office
for the parcel map; the state data center for census numbers, the planning
board or select board for zoning; the soil conservation service for soils classifications, the county extension office for resource inventories, and USGS for
topographic maps. A comprehensive spatial database that integrates digital
maps, descriptive attributes, digitized photography, and imagery with a GIS
can dramatically improve the productivity of planners by providing them, in
essence, with a centralized "one-stop" data depository.

7.6.4. Training and computational overhead

The integration of remote sensing into planning has until recently been limited to large, well-funded organizations due to the cost of establishing and
maintaining an image processing platform. Visual and computer analysis of
aerial and orbital imagery is a specialized domain of activity that is generally
considered to be heuristic and based in experience (Millette, 1989). Professionally trained planners generally receive little training in areal photographic
interpretation, much less image processing. Therefore, unless a planning organization is large enough to dedicate a staff position to a specialist, remote
sensing has been utilized only when large projects could fund consultants.
Early image processing systems were either large and expensive software
products that ran on mainframes or dedicated integrated minicomputer hardware and software environments. In either case, few planning organizations
had the capital and human resources to invest in such technology. However,
with the rapid growth in geoprocessing technology that has taken place over
the past decade and the increased power of computers at decreasing prices,
opportunities for image processing platforms exist even in the most mod-
estly funded planning organizations. For example, image processing software
packages costing a few hundred dollars can run on pentium class general purpose microcomputers costing less than $3,000. Larger planning organizations

with more production-oriented image processing requirements can find high productivity in the workstation environment with hardware and software combinations in the $20,000 to $50,000 range.

Although few planners are likely to have any in-depth training or experience in image analysis, many planning organizations are developing new staff positions for GIS technicians and application specialists (Gallagher, 1992). Since GIS planning staff are already familiar with hardware, software, and geographic data structures, image processing can be considered just an extension of their existing GIS activities. As such, with relatively modest training focused on a quantitative understanding of image characteristics, spectral enhancements, and topographic illumination effects, together with statistical training and classification approaches, planning GIS staff can be expected to develop a functional remote sensing capacity. For example, GIS staff at four Vermont regional planning commissions were given a three-week training course that allowed them to develop regional land cover classifications with accuracies ranging from 88% to 93% (Millette, 1992). Given that most planning organizations give technical staff members opportunities for on-going training through seminars and workshops, it is reasonable to expect that remote sensing expertise should continue to grow with ongoing training and experience.

The amount of computational overhead created by a remote sensing platform in a planning agency is modest for those organizations that already have GIS programs. Because planning organizations cannot function today without basic computer tools such as word processing, spread sheets, database management, and traffic models, most will already have a microcomputer platform that can be used on a part-time basis for image processing. Since most planning agencies only use remote sensing for semiannual land cover updating, investing in dedicated hardware, software, and a staff position for remote sensing is difficult to justify for the average municipality. However, making use of existing hardware and personnel together with a modest investment in software and training may be cost effective for some planning agencies. Planning organizations that currently do not have a GIS platform or geoprocessing experience would be prudent to continue using individual consultants or value-added image processing firms on a project basis rather than investing in digital remote sensing infrastructure that would only be used on an occasional basis. Indeed, it is important to note here that the more traditional photogrammetric engineering firms that city agencies already contract with are beginning to branch out and offer both image processing and GIS-related services.

7.7. Standards

Standards for spatial data in the planning environment are critical to the accuracy, effectiveness, and productivity of planners and planning assessments. The spectrum of considerations with respect to standards include: data accuracy and quality, database administration, geodetic control, telecommunications, data exchange and interchange, and documentation. If you consider that activities of spatial data processing involve the commingling of geocoded data taken from a variety of different sources, as the quality of those data vary, so docs the spatial database being developed. Although spatial data standards represent too large a spectrum of concerns than can be treated here in any detail (see Maguire, Goodchild, and Rhind, 1991 for more complete treatments), it is useful (albeit artificial) to classify them into one of two classes: (1) those that affect accuracy and error and (2) those that affect handling and processing.

Standards that affect accuracy and error include positional accuracy of map products based on identifiable cartographic features (ASPRS, 1989) used as inputs to and outputs from planning assessments; attribute accuracy of categorical data such as land use classes generally assessed by spatial sampling and represented in a error matrix (Congalton, 1988; Dicks and Lo, 1990); and completeness and consistency, such that the number and character of geographic objects in the environment are represented in the spatial database (Chrisman, 1991). With regards to positional accuracy, U.S. National Map Accuracy Standards require maps published at scales of 1:20,000 and larger to have no more than 10% of points tested greater than $1/30$ inch from their true location. For elevation data, the standard is that no more than 10% of sampled points can be in excess of $1/2$ the contour interval from the location's true elevation. At scales smaller that 1:20,000 this distance is decreased to $1/50$ of an inch. Attributes from classified satellite data generally need to have an average (diagonal) accuracy of 85% to be considered acceptable. Although there are no universally accepted prescriptions for consistency and completeness checking, a commonly used approach is to match records in the spatial database against some external reference list considered to be without errors.

Standards that affect data handling and processing include map coordinate systems, spatial and attribute data structures, coding schemes for attributes, and documentation procedures for data layer development. Most planning-related assessments are done with data that have been captured or compiled by some other administrative or governmental agency, rather than produced in-house from scratch. One of the most important considerations in pulling together data from dispersed sources is its geocoding and registration to some common map base. Without careful consideration to procedures for putting data onto a map, important topological relationships such as adjacency and proximity may not be preserved.

Municipal and regional planning organizations generally use either the local State Plane or UTM coordinate systems as coordinate bases for spatial database development. A common map base provides the spatial framework to ensure that data from dispersed sources will ultimately "fit" together. For example, states such as North Carolina and Vermont have encouraged through legislation the use of orthophoto control of GIS databases. The Vermont State GIS requires that all spatial data captured by planning organizations be transferred to Vermont's 1:5,000 orthophotos registered to the Vermont State Plane coordinates before being digitized in order to conform to state data standards (Millette, 1990).

The use of standardized attribute coding schemes and documentation procedures are essential to efficient use and quality assurance of spatial databases. There is presently a significant effort being marshaled into cooperative spatial database development at the international, national, regional, state, and local levels. The impetus for these cooperative efforts is that, due to its cost, spatial data can no longer be considered a parochial resource, but a shared asset to be developed, capitalized, and used by the broadest community possible. Standardized and well-documented coding schemes are essential to support proper use and integration of spatial data. For example, without standardized (and adhered to) coding schemes for descriptive attributes, land cover classifications developed in one region of a state may not be compatible with those developed in another. Although most states in the United States have their own statewide land use/land cover coding systems, most are based in some measure on the one developed by the USGS (Anderson, Hardy, and Roach, 1976).

Documentation procedures are generally considered a data management function and are essential if users are to have confidence in the data and are to be able to use it appropriately (e.g., they cannot assume greater accuracy than is justified). Data documentation should include:

- a comprehensive data dictionary with descriptions of all data elements and codes for each GIS data layer (both geographic and attribute);
- a detailed presentation of the classification scheme, including concepts and organization; and
- a complete listing of sources and formats for all geographic and attribute elements in the GIS, including data capture and conversion procedures, processing tolerances, and genealogy.

Figures 7.3 and 7.4 illustrate examples of a GIS data layer documentation and data layer history for a portion of the Vermont GIS database. Note that the documentation file carries a detailed amount of information concerning the origin and development of the GIS layer whereas the history file provides a record of all changes made to the layer over its life. Figure 7.5 illustrates the quality assurance documentation scheme for one segment of the street centerline database in the Clark County, Nevada GIS. Note that the QA value is comprised of two digits characterizing (1) the source used to create the data and (2) the original cartographic scale of each GIS data source. Clark County reduces both the risks and misuses of a public domain database by attributing accuracy assessments for each data element.

Figure 7.3. Data layer documentation file for Woodstock land cover layer
(courtesy of author)

```
COVER-NAME             = WOODLC
COVER-CONTENT          = WOODSTOCK LAND COVER
SOURCE-AGENCY          = TRO REGIONAL COMMISSION
DATA-SOURCE            = ORTHOPHOTO 144120
SOURCE-MAP&DATE        = VERMONT ORTHOPHOTO 1989
SOURCE-SCALE           = 1:5000
SOURCE-MEDIA           = PAPER
SOURCE-ACCURACY        = 10'
SOURCE-RESOLUTION      = 3'
COMPILATION-DATE       = 1989
DIGITIZING-DATE        = 6/10/91
DIG-PERSON             = T.L. MILLETTE
DIG-DEVICE             = GTCO 5A-DIGI-PAD
DIG-RESOLUTION         = .001"
METHOD-QUAL-ASSR       = FIELD EXAMINATION
MAP-PURPOSE            = GENERAL INVENTORY
PROJECTN-&-UNITS       = STATE PLANE-METERS
DATUM                  = NAD 1927
DATA-TYPE              = POLY
GIS-SOFTWARE           = PC ARC-INFO
SOFTWARE-VERSION       = 3.3D
RMS-TOL                = 1.1
DANGLE-LENGTH          = 25
LINE-ACCURACY          = 90% W/IN .01'
MAP-EDITION            = DRAFT
DIG-ORGANIZATION       = TRO REGIONAL COMMISSION
CONTACT-PERSON         = E.EDELSTEIN
STREET-ADDRESS         = KING FARM
CITY-STATE-ZIP         = WOODSTOCK, VT 05091
PHONE                  = (802) 457-3188
NOTES 1                = SATELLITE DATA PILOT PROJECT
NOTES 2                =
```

7.8. Future trends in planning use of integrated spatial data

The 1980s were marked by increases in development pressures and local responsibilities for most planning organizations without commensurate increases in funding. Early in the 1990s the unprecedented volumes of spatially referenced digital data has offered planners opportunities to formulate cost-effective analytical approaches to meeting the continually increasing planning mandate. Digital data currently available from the National Digital Cartographic Data Base (NDCDB), which is part of the new National Spatial Data Infrastructure (NSDI), and a host of new commercial sources offer planners unparalleled access to street network, socioeconomic, demographic, and marketing information. Commercial data firms provide value-added enhancements to the format, coverage, and quality of information, utilizing data quality as a basis of their pricing structures (Castle, 1993). Digital data on regional infrastructure proliferate as utility companies automate their facilities and distribution networks. These digital infrastructure data encompass a full range of

Figure 7.4. Data layer history file showing hardwood category update
(courtesy of author)

```
DATAFILE-NAME   =   WOODLC.HST
DATE            =   9/10/91
OPERATOR        =   T.L.MILLETTE
MAP             =   VT ORTHOPHOTO 144120
PROCESS         =   ARCEDIT
RMS             =   .002
NOTES           =   HARDWOOD UPDATE DUE TO HARVEST
```

Figure 7.5. QA coding scheme for street centerline Records in Clark County, Nevada (courtesy of author). Examples of QA values: 32 = digitized input from 1:1,200 scale orthophotography; 43 = input from coordinate geometry file at 1:2,400 scale

Spatial scale source accuracy	Original document source value
0 Reserved	0 Ortho(edges defined)
1 Control point	1 Registered to parcel
2 Design drawing	2 Ortho(edges undefined)
3 ≤ 1:1,200	3 COGO data
4 ≤ 1:2,400	4 Adjusted COGO data
5 ≤ 1:4,800	5 Reserved
6 ≤ 1:9,600	6 Reserved
7 ≤ 1:24,000	7 Digital line graph (DLG)
8 ≤ 1:50,000	8 Interpreted from DIME
9 > 1:50,000	9 Other

planning elements for utilities, gas, electrical, transportation, and communications. Satellite data are increasingly being used in lieu of traditional aerial photography for a variety of applications as finer spatial resolutions become available. Remote sensing data provide planners with accurate and timely information for a variety of urban topics, such as land cover, change detection, and facility siting. Issues of positional accuracy and georectification of digital data are benefiting from the advances and affordability of GPS technology. GPS is providing a powerful and economical geocoding tool to upgrade and enhance digital data sets.

The advent of spatial data standards for information interchange and exchange, as supported under the federal government's SDTS program, will improve planners' access to inexpensive data repositories. Commercial data exchange standards continue to emerge as market forces encourage information sharing and cost reduction. Issues of data integrity and quality assurance in large multiuser GIS databases will continue to stimulate metadata research for cataloging and maintenance of data quality. Improvements in GIS hardware and software tools enhance planners' ability to perform a variety of sophisticated analyses. Powerful workstations in the $10,000 – $20,000 price range are operating at an ever-increasing number of MIPS (million instruction per second). Affordable workstations will promote the use of desktop analyses in planning organizations. Digital data storage media that handle gigabytes and terabytes of data are expected to keep pace with the bulk data processing speeds required in large spatial databases. A software evolution, fueled by increasingly sophisticated user demands, offers promise for eliminating the presently complex user interfaces common to many GIS and remote sensing software packages. Strategies for improving software tools include simplifying and standardizing graphic user interfaces (GUIs) and integrating artificial intelligence (AI) components and front-ends. At the same time, geoprocessing-literate planners are being developed from a plethora of GIS courses offered by universities and systems vendors. Advances in digital mapping software are improving planners' understanding of geographic data and promoting use of the multimedia visual images. Statistical analyses viewed in a 3D display with time-series projections provide improved visualization of planning assessments prior to land use policy implementation. Satellite images draped over 3D terrain elevation models provide planners with better perspectives of the landscape at regional and local scales. Image-based information systems are projected to increase the use and benefits of integrating remotely sensed data and GIS formatted data into the planner's workplace (Millette, 1992).

The neglect of the 1980s has exacerbated urban decay in the 1990s. Changes in municipal service demands resulting from urban decay and suburban sprawl will continue to bring a host of challenging problems to the planner's plate. The growth in environmental protection regulations will require specialized training for planners to address nuances of complex federal, state, and local mandates for clean air, clean water, and safe handling of hazardous and toxic materials. Adding to a planner's mission will be the difficult task of siting waste handling facilities, which often prove politically unacceptable in anyone's backyard. Traffic and transportation planning will require attention to a wide spectrum of sensitive environmental topics. A renewed focus on health care and education will likely add to the regional and urban planner's workload. Master planning reports and local development studies will need to incorporate the health, safety, and educational impacts as part of each project and policy evaluation. Cuts in government support combined with increased demands for "omniscient planning" will provide little latitude within a planner's resource budget when developing alternative planning scenarios. At the same time, planners will need to adopt improvements in the analytical methods used to provide decision makers and community watch groups with comprehensive policy options when confronting these future challenges. Policy makers are making increased demands on planners for information to manage regional and urban change. Increased use of decision support systems (DSS) and AI linked to GIS databases will be important in the future to meet the demands placed on planners.

Planners' use of spatial analytical tools will be required to meet a host of new legislative demands. Community development support funding from federal programs will continue to be tied to compliance of increasingly complex environmental and social laws. Planners will need to use GIS for a variety of programs to cost-effectively monitor compliance with new regulations originating under programs such as the national pollution discharge elimination systems (NPDES) and the American Disabilities Act (ADA). The proactive planning community must be aware of these trends and harness the power of spatial analysis systems with remote sensing data inputs. The planner's solutions to the challenges of the future appear inextricably linked to the resources of regional and local GIS databases and the improved applications available through integrated tools of GIS and remote sensing.

8

Global change research and geographic information systems requirements

Ghassem R. Asrar

8.1. Introduction

The inhabitants of Earth have become a dominant force in competing and interfering with natural processes that control the balance and functioning of our life support system, the Earth system. Reconciliation of this problem and our desire to improve our global standard of living are not mutually compatible, thus placing us and the future generations in a very difficult position of having to choose between preserving the environment or continuing to promote industrial and agricultural development for short-term gains at any cost. There are two possible methods of resolving this conflict: (1) rely on our ability to adapt to environmental changes regardless of their heritage and (2) mitigate the currently known factors and/or processes that are causing and/or contributing to environmental changes. In both cases, one has to have access to necessary information in a reliable and timely manner, and in suitable format(s) for facilitating the making of decisions and/or establishment of environmental policies.

There is global recognition of environmental issues and concerns, and many concerted efforts at both the national and international level are underway to provide, for the first time in history, tremendous amounts of data and information on Earth's environment during the next century. Some of the challenges facing natural and social scientists, however, are how to: (1) combine and store data collected through diverse methods and sources; (2) merge and analyze these data; (3) couple these data with numerical simulation models developed for predicting global climate changes and assessing their impact; and

(4) display and present the results of the analyses in a comprehensible manner to the rest of scientists, policy makers, and the public.

The rapid development and evolution of computer hardware and software during the past decade offers the Earth scientist valuable tools to cope with these challenging issues. Notable among these developments are the geographic information systems, which are ideal for merging, analyzing, and displaying data collected from a variety of sources.

The purpose of this chapter is to present an overview of the planned remote sensing research efforts at the national and international levels and to outline some of the unique challenges that the information system scientists will be facing in helping to provide the necessary tools to archive, distribution, analyze, and display this wealth of environmental data during the next two decades.

8.2. Global change research program

During the 1980s the general scientific awareness of the need to conduct multidisciplinary research and field experiments continued to grow. Coordinated field studies such as the International Satellite Cloud Climatology Project (ISCCP) and the International Satellite Land Surface Climatology Project (ISLSCP) were defined and conducted during this period. The National Aeronautics and Space Administration (NASA), initiated a series of studies during this period to define a comprehensive space-based global observing system to enable an integrated approach to studying the Earth's atmosphere, oceans, and land surfaces. The objective of these studies was to focus not only on long-term monitoring and a detailed study of each of these components separately, but also to assess and understand the couplings and exchange processes that govern the interactions and interconnections among them. These deliberations resulted in definition of the NASA Mission To Planet Earth (MTPE) program in 1982.

The national and international Earth science communities also engaged in a series of planning activities under the auspices of the World Climate Research Program (WCRP), International Geosphere-Biosphere Program (IGBP), and the International Council of Scientific Unions (ICSU) to define a series of multinational and multidisciplinary research programs and studies during this period. For example, the Global Energy and Water Experiment (GEWEX), World Ocean Circulation Experiment (WOCE), Tropical Ocean-Global Atmosphere (TOGA), and a number of other initiatives resulted from these activities. Within the United States, government agencies formed the Committee on

Earth and Environmental Sciences (CEES) to develop the plan for a national Global Change Research Program (GCRP). The Intergovernmental Panel on Climate Change (IPCC) was formed at the international level to carry on the scientific assessment and to define and to prioritize the scientific and policy issues related to global change. The scientific issues and priorities identified through these efforts were adopted by NASA as a focus for the MTPE, which is NASA's contribution to the national and international GCRPs.

The United States GCRP (USGCRP) will develop and use space- and ground-based measurements to provide the scientific basis for understanding global change. NASA's contributions to USGCRP include ongoing and near-term satellite missions, new programs under development, planned future programs, a continuing basic research program focused on process studies and modeling, and a comprehensive data and information system. The space-based components of the mission consists of a constellation of satellites that will monitor the Earth from space. But, space-based research and monitoring are not sufficient; a comprehensive data and information system and a community of scientists performing research with the data acquired and conducting extensive ground-based research campaigns are all important components of the MTPE. The commitment to provide the necessary tools for archiving, analysis, distribution, and display of the Earth science data by the research community is critical for the success of the program.

The ultimate goal is to establish the sound scientific basis for an intense study of planet Earth to help make sound environmental policy decisions during the twenty-first century. To put in proper perspective the need for the data analysis tools, it would be helpful to present first the objectives and a brief description of the components of MTPE, then to focus on description of the data and information analysis tools needed by the mission. It is in this context that the GIS requirements and challenges will be discussed.

8.3. Mission objectives

The four distinct objectives of the USGCRP, and thus the MTPE program, are:

- Documenting global change through the establishment of an integrated, comprehensive, and long-term program of observing and analyzing Earth system changes on regional and global scales, including data and information management;
- understanding the key Earth system processes through a program of focused studies to improve our knowledge of (1) the physical, chemical, biological,

geological, and social processes that influence and govern the functioning of the Earth system and (2) the effects of global changes on natural systems and human health and activities;

- predicting global and regional environmental changes through the development and application of integrated conceptual and predictive Earth system models; and
- assessing and synthesizing the state of scientific, technical, and economic knowledge and implications of global changes to support national and international policy-making activities and to provide guidance for determining research priorities of the USGCRP.

All four elements must work together interactively for the USGCRP to succeed (Figure 8.1). It is obvious that data and information management is a critical component of both USGCRP and MTPE.

MTPE combines the development of a series of instruments and spacecraft (space-based components), which will provide the necessary observations with a comprehensive data and information system, scientific research and

Figure 8.1. Components of the U.S. global change research and mission to planet Earth programs

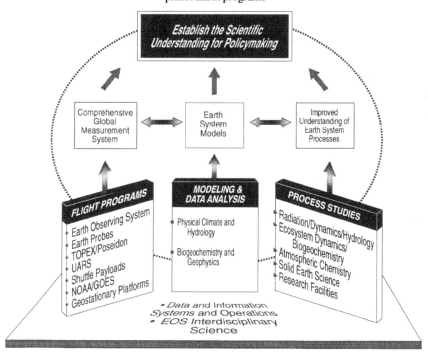

analysis, education, and the national and international coordination (ground-based components) to accomplish these objectives.

8.4. Space-based components

The space-based component of MTPE consists of satellites that are currently operational and those either under development or planned for the future to fly in a variety of orbits (polar, inclined, and geostationary). There is no single orbit that permits the gathering of complete space-based observations of Earth processes.

8.4.1. Current satellites

The currently operating satellites include:

- The *Total Ozone Mapping Spectrometer* (*TOMS/Meteor-3*), launched in 1991 on a Russian meteorological satellite to measure the concentration of ozone in the atmosphere. Measurements from this instrument will supplement and continue the measurements from the *TOMS* instrument on the *NIMBUS-7* satellite from 1978 until May 1993.
- *Upper Atmospheric Research Satellite* (*UARS*), launched in 1991. It carries instruments that have contributed to studies of the distribution of stratospheric ozone and other trace gases and furthered our understanding of the interaction among dynamic, radiative, and chemical processes in the Earth's stratosphere and mesosphere.
- The *Earth Radiation Budget Experiment* (*ERBE*), a program launched in 1984 to investigate the role of clouds, solar radiation, and aerosols in heating and cooling of the atmosphere. It has also contributed to measurements of ozone concentration. *ERBE* data has been supplemented with data from instruments launched in 1984 and 1986 on various environmental satellites in different orbits.
- *Ocean Topography Experiment* (*TOPEX/Poseidon*), a joint effort with France launched in August 1992 to study ocean circulation and its influence on air-sea exchanges of heat, mass, and momentum.
- *Laser Geodynamic Satellite* (*LAGEOS-2*), a joint effort with Italy launched in 1992 to monitor crustal motion and variations in Earth's rotation.

In addition to these dedicated research missions supported by NASA, Earth scientists conduct research using data from operational satellites operated by the U.S. National Oceanic and Atmospheric Administration (NOAA), its

Table 8.1. *Space-based Earth science data from non-NASA sources*

Non-NASA data sources	Measurement objectives
NOAA-9 through-*J* (U.S. – operating)	Visible and infrared radiance/reflectance, infrared atmospheric sounding, and ozone measurements
Landsat-4/5 (U.S. – operating) *Land Remote Sensing Satellites*	High spatial resolution visible and infrared radiance/reflectance
DMSP (U.S. – operating) *Defense Meteorological Satellite Program*	Visible, infrared, and passive microwave atmospheric and surface measurements
ERS-1 (ESA – operating) *European Remote Sensing Satellite-1*	C-band Synthetic Aperture Radar (SAR), microwave altimeter, scatterometer, and sea surface temperature
JERS-1 (Japan – operating) *Japan's Earth Resources Satellite-1*	L-band SAR backscatter and high spatial resolution visible and infrared radiance/reflectance
ERS-2 (ESA – January 1995) *European Remote Sensing Satellite-2*	Same as *ERS-1*, plus ozone mapping and monitoring
Radarsat (Canada – March 1995) *Radar Satellite*	C-band SAR measurements of Earth's surface (joint U.S./Canadian mission)
NOAA-K through-*N* (U.S. – September 1994 on)	Visible, infrared, and microwave radiance/reflectance; infrared atmospheric sounding; and ozone measurements
ADEOS (Japan – February 1996) *Advanced Earth Observing System*	Visible and near-infrared radiance/reflectance, scatterometry, and tropospheric and stratospheric chemistry (joint with U.S. and France)

counterpart environmental monitoring agencies around the world, the U.S. Department of Defense (DOD), the U.S. *Landsat*, and the French *Systeme pour l'Observation de la Terre* (*SPOT*) satellite. Brief descriptions of data from the non-NASA sources are given in Table 8.1.

8.4.2. Satellites under development

Satellites under development by NASA include those missions that are focused on the specific requirements of individual Earth science disciplines, referred to as Earth Probes, and missions focused on interdisciplinary studies of the Earth as an integrated system through the Earth Observing System (EOS) program.

NASA's Earth Probes program plans to launch a satellite every two to three years. Earth Probes missions under development include:

- The *Sea Wide Field Sensor* (*SeaWiFS*), an ocean color sensor to study ocean productivity and interactions between marine ecosystems and the atmosphere, will be launched in 1995.
- Additional copies of the *TOMS* instrument will fly on a NASA explorer-class satellite in 1994 and on Japan's *Advanced Earth Observation Satellite* (*ADEOS*) in 1996 to ensure the continuity of vital global ozone measurements.
- *Radarsat* is a Canadian satellite for which NASA supplied a launch in 1995. The satellite carries a C-band synthetic aperture radar, which is valuable for study of the Earth's polar regions.
- The *NASA Scatterometer* (*NSCAT*), an instrument to measure the oceans' surface stress and to determine wind patterns and their contribution to air-sea interactions, will also be launched on Japan's *ADEOS* satellite in 1996.
- The *Tropical Rainfall Measurement Mission* (*TRMM*) is a joint program with Japan, scheduled for launch in 1997. *TRMM* will make extensive measurements of precipitation in tropical regions, which is the heat engine for global climate and cannot be time-sampled adequately from polar orbit.
- NASA Spacelab Series includes the *Shuttle Imaging Radar-C* (*SIR-C*), which is a joint mission with Germany and Italy; *Scanning Solar Backscatter Ultra Violet Experiment* (*SSBUVE*); *Lidar In-space Technology Experiment* (*LITE*); and the *Atmospheric Laboratory for Applications and Science* (*ATLAS*). These missions are advanced spaceborne instrument development campaigns scheduled for deployment on multiple NASA space shuttle missions in 1994 and beyond.

The centerpiece of NASA's contributions to MTPE is the Earth Observing System, a series of satellites in polar and mid-inclination orbits for long-term global observations of the Earth's atmospheric chemical and physical processes, terrestrial and oceanic ecosystem processes, oceanic circulation, and solid Earth processes. Combined with polar-orbiting and mid-inclination platforms from Europe and Japan, EOS will form a comprehensive international system. The EOS program together with the international contributions will provide space-based observation series of at least fifteen years. Scientists will be able to access data and information obtained by EOS instruments at many levels of detail, covering all of the major Earth system processes over a wide range of spatial, spectral, and temporal domains. Table 8.2 describes the measurement objectives of the international Earth observing satellites and their scheduled launch dates.

8.4.3. Future satellites

EOS is a long-term mission that will continue to evolve over the next twenty years. The scientific and technological lessons learned after the launch of the first series of satellites in the late 1990s will be used to design and develop new sensors and satellites to provide continuity and new measurement capabilities. This dictates EOS in all aspects to be an evolutionary program.

The Earth Probes program will also continue to support the need for continuity of some existing measurement capabilities. It will also fulfill additional requirements that either have already been identified (e.g., measurements of gravity and magnetic fields and land surface topography) or will be identified by the Earth science community in the future.

The MTPE plan also calls for a series of geostationary satellites to provide continuous observations of the Earth that permit intensive study of daily variability and changes in the Earth system over extended time periods. Much as geostationary weather satellites track storm systems today, these satellites will monitor dynamic short-term phenomena (e.g. floods, fires) that cannot be observed from polar or low-inclination orbits. The scientific objectives for the geostationary satellites focus on increasing our understanding of short-term processes needed to build mathematical models that adequately simulate the Earth system.

Unlike the EOS and Earth Probes Programs, geostationary satellites will require the design and development of new or improved advanced technology. NASA plans to proceed with definition activities during the 1990s, leading to the launch of the first geostationary satellite after the turn of the century. Experience with these geostationary research satellites will contribute to the planning and development of the next generation of geosynchronous weather satellites.

In addition, NASA has established memoranda of understanding to share the data from future space-based observations of the European Space Agency (ESA), Canada Space Agency (CSA), and the National Space Development Agency (NASDA) of Japan in return for having access to NASA-sponsored missions. Details of such agreements are being negotiated by a number of existing international coordination working groups and committees. It is hoped that these efforts will lead to a coordinated national and international program of Earth observations, which will provide a comprehensive global data and

Table 8.2. *The measurement objectives of the International Earth Observing System (IEOS)*

IEOS elements	Measurement objectives
EOS-AM Series (NASA – June 1998 on) *Earth Observing System Morning Crossing* *(Descending)*	Clouds, aerosols, and radiation balance; characterization of the terrestrial ecosystem; land use, soils, and terrestrial energy/moisture; tropospheric chemical composition; contribution of volcanoes to climate; and ocean primary productivity (includes Canadian and Japanese instruments)
EOS-COLOR (NASA – August 1998) *EOS Ocean Color Mission*	Ocean primary productivity (afternoon crossing)
ENVISAT Series (ESA – September 1998 on) *Environmental Satellite*	Environmental studies in atmospheric chemistry and marine biology; continuation of *ERS* mission objectives
ADEOS II (Japan – May 1999) *Advanced Earth Observing Satellite II*	Visible, infrared, and microwave radiance/reflectance; scatterometry; microwave atmospheric sounding; and tropospheric chemistry (includes U.S. instruments)
HIROS Series (Japan – February 2002 on) *High Resolution Observing System*	Visible, infrared, and microwave radiance/reflectance; SAR; infrared and laser atmospheric sounding; tropospheric and stratospheric chemistry; radar; precipitation and altimetry (may include French and U.S. instruments)
EOS-AERO Series (NASA – 2000 on) *EOS Aerosol Mission*	Distribution of aerosols and greenhouse gases in the lower stratosphere (spacecraft to be provided through international cooperation)
EOS-PM Series (NASA – 2001 on) *Earth Observing System Afternoon Crossing (Ascending)*	Cloud formation, precipitation, and radiative properties; air-sea fluxes of energy and moisture; vegetation and ocean productivity; snow cover and sea-ice extent (includes European instrument)
METOP Series (ESA – June 2000 on) *Meteorological Operational Satellite*	Operational meteorology and climate monitoring, with the future objective of operational climatology (joint with EUMETSAT and NOAA)
EOS-ALT Series (NASA – 2002 on) *EOS Altimetry Mission*	Ocean circulation, land topography, and ice sheet mass balance (may include French instruments)
EOS-CHEM Series (NASA – 2003 on) *EOS Chemistry Mission*	Atmospheric chemical composition, aerosols, and solar radiation (includes Japanese and joint U.S.-U.K. instruments)

information system for studying the Earth as an integrated system over the next several decades.

8.5. Ground-based components

The ground-based components of the MTPE consist of a comprehensive and an end-to-end system of acquisition, processing, and distribution of remotely sensed observations provided by the space-based components and in situ observations required to aid calibration, validation, and/or interpretation of these data by Earth scientists. MTPE also supports scientific investigation and analysis of these data by Earth scientists. Training and education of the next generation of Earth scientists through a wide range of educational activities is another essential component of the MTPE (NASA, 1993).

8.5.1. *In situ observations*

The ground-based component of the MTPE consists of the in situ observations required to calibrate, validate, and interpret remotely sensed observations provided by the space-based component. Some of these in situ observations will be obtained by NASA-funded investigations over selected regions of the Earth, but NASA by itself cannot support a comprehensive global in situ observation program. To fulfill this requirement, NASA has signed a number of memoranda of agreement with other U.S. federal agencies participating in USGCRP that have the in-house expertise and ongoing research and operational programs to provide the necessary measurements in return for having access to the space-based observations. Examples include the Department of Energy (DOE) Atmospheric Radiation Measurement (ARM) program; the National Science Foundation (NSF) Long Term Ecological Research (LTER) sites and the USGS and NOAA operational networks of atmospheric, oceanic, and terrestrial monitoring stations. Similar agreements have been or are being negotiated with some foreign countries for their participation in MTPE through the WCRP, the IGBP, the Global Climate Observing System (GCOS), the Global Oceans Observing System (GOOS), and the Global Terrestrial Observing System (GTOS).

8.6. Science investigations

MTPE investigations support development and operation of remote sensing instruments and the conduct of interdisciplinary investigations using data from these instruments. These investigations are conducted by scientists from academia, private industry, other Federal agencies, and NASA field centers who are selected through peer evaluation processes. The main objective is to understand, model, and assess the role of fundamental biological, chemical, and physical processes that govern and integrate the Earth system. These include:

- Hydrologic processes, which govern the interactions of land and ocean surfaces with the atmosphere through transport of heat, mass, and momentum. Evaporation, transpiration, ocean circulation, and precipitation processes and their contribution to the atmospheric forcing and feedback mechanisms is a central issue.
- Biogeochemical processes, which contribute to the formation, dissipation, and transport of trace gases and aerosols and to their global distribution.
- Climatological processes, which control formation and dissipation of clouds and their interaction with solar radiation. This involves understanding the role of clouds, radiation, water vapor, and precipitation in climate systems and their local, regional, and global distributions.
- Ecological processes, which are effected and/or will affect the global change, and their response to such changes through adaptation.
- Geophysical processes, which have shaped or continue to modify the surface of the Earth through tectonics, volcanism, and the melting of glaciers and sea ice.

Developing Earth system models that represent the interactions among complex Earth system processes at all spatial and temporal scales is an essential component of MTPE. Such models will help us understand how Earth functions as an integrated and coupled system, and they serve as a tool for assessing the impacts of global climate changes on agricultural and industrial developments. To derive full benefits from our investments in MTPE we must assess the economic and social impacts of global climate change with the help and participation of social scientists and policy makers.

8.7. Education and training

The goal of the MTPE education program is to enhance awareness, interest, and knowledge of Earth system science by teachers and students (preschool to graduate) and the public. This goal is accomplished through the projects that empower teachers with creative approaches to teaching Earth system science within existing curriculum; enhance student experiences, provide financial support for undergraduate and graduate student research and training, and stimulate cooperative activities with universities and other national and international agencies.

For example, the Graduate Student Fellowship in Global Change Research was established in 1990 to support graduate students pursuing Ph.D. degrees in Earth science. The main objective of this program is to train the next generation of scientists and engineers to help analyze, interpret, and manage the wealth of data and information generated by MTPE. The program was envisioned to scale up to fund 150 graduate students prior to the launch of the first EOS satellite in 1998 and to remain at that level during the life of the MTPE program. Due to overwhelming response over its first four years, the program has already awarded more than 160 fellowships during the past two academic years, thus exceeding its intended goal. NASA intends to continue supporting this and a large number of other educational projects, which are described briefly in the *MTPE Catalog of Education Programs and Activities* (NASA, 1993).

8.8. Data and information system

The USGCRP will generate massive quantities of highly diverse data and information in support of its stated scientific objectives. These data are provided through the existing programs and archives in addition to those from new initiatives designed to provide the required additional data and information. The basic challenge facing USGCRP is how to manage the large volume of data resulting from these diverse initiatives in an effective manner useful to the global change research community.

Since 1957, a large number of studies conducted under the auspices of the U.S. National Academy of Science (NAS) and other federal agencies have outlined the problems and made specific recommendations for improved data and information management systems. In 1991, the NAS recommended the coordinated development of an interagency global change data and information system (GCDIS). A plan has been developed by U.S. federal agencies to collaborate with each other, with academia, and with the international community to enhance the collection, archiving, and distribution of global change data and information (CEES, 1992). The EOS Data and Information System (EOSDIS) will be the NASA component of the GCDIS. NASA has also agreed to secure the necessary funds for establishing interoperability among the GCDIS nodes on behalf of the entire USGCRP.

The EOSDIS will receive, process, catalog, archive, and distribute all EOS data from existing and future Earth remote sensing missions and will be interoperable with other components of GCDIS and the data and information systems of the international partners. EOSDIS is a comprehensive and end-to-end data and information system. It must perform a wide variety of functions, supporting individuals located in various organizations and nations, by conducting several distinct types of activity, including:

- mission planning, scheduling, and control;
- instrument planning, scheduling, and control;
- resource management;
- communications;
- computational facilities at investigator sites;
- generation of standard data products;
- generation of special data products;
- archiving of data, products, and research results;
- data and information cataloging, searching, browsing, and ordering;
- effective distribution of all information holdings; and,
- user support.

These groupings of functions and activities together with their interrelationships are illustrated in Figure 8.2. Multiple boxes illustrate that EOSDIS must have distributed capabilities. For example, data used in research, algorithm development and maintenance, and data product generation, archiving, and distribution are carried out in many different locations (i.e., investigator's institutions), whereas mission planning, scheduling, and control take place at one site (i.e., the EOS Operations Center).

EOSDIS will have a distributed, open-system architecture. This allows for the distribution of EOSDIS elements to various locations (Figure 8.3) to take best advantage of different institutional capabilities and science expertise (NASA, 1992). Although EOSDIS is physically distributed, it will appear as a single logical entity to the user through the network connections. EOSDIS will consist of an EOSDIS Core System (ECS) to provide centralized mission and instrument command and control functions and distributed (but common) product generation, archive, and information management functions. Capabilities also exist outside of the core, including site-unique extensions to core capabilities, computing facilities for EOS researchers, and so on. Figure 8.3 shows the EOSDIS architecture and the diversity of its component elements, most of which will be geographically distributed. These elements are described in detail in the EOSDIS brochure (NASA, 1992).

The GIS approach to data management has risen in popularity to become a key driver for EOSDIS interoperability requirements. Although the issues and challenges related to coordination and development of data and information

Figure 8.2. A conceptual outline of the integrated Earth Observing System Data and Information System (EOSDIS) and its functional elements

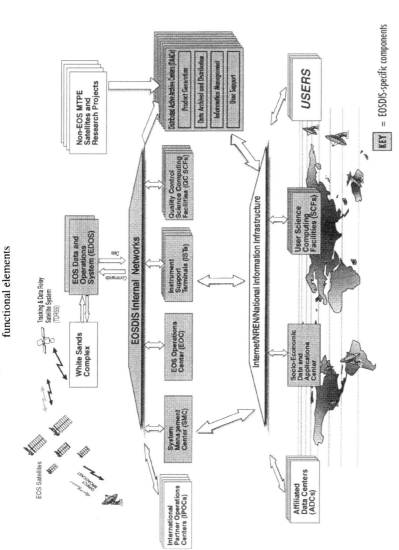

systems at the national and international levels require more detailed discussions than are appropriate here, it is important that we continue to study the evolution of GIS and their potential contributions to the data and information analysis objectives of the MTPE and USGCRP.

8.9. Evolution of geographic information systems

Geographic information systems allow integration of datasets acquired by in situ and remotely sensed observations of the Earth system processes within the context of relatively constant geographical features. The spatial and temporal diversity of these observations can provide a rich source of information but may be difficult to analyze and display in a meaningful manner. Geographic information systems may provide the means for such analysis and display. A comprehensive review of GIS was presented recently by Estes and Star (1990). The brief evolutionary history of GIS presented in the following will provide a basis for establishing the future requirements of the Earth system scientists.

Figure 8.3. The geographical distribution of the distributed active archive centers of the EOSDIS and their Earth science discipline orientations.

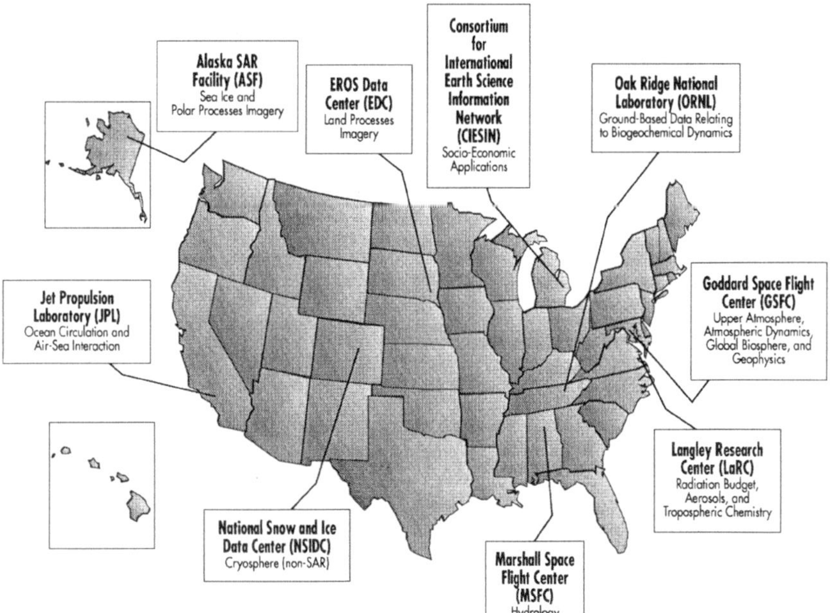

The development of GIS started in the mid-1960s by using computers and computer graphics to do tasks such as mapping, in the absence of interactive spatial or statistical analysis tools. By the early 1970s, more sophisticated GIS with statistical and spatial analysis tools and three-dimensional graphics were being developed. These systems did not gain broad acceptance for scientific data analysis and decision making until late 1970s. The introduction of personal and minicomputers and object-oriented programming during the 1980s decentralized the use of GIS and resulted in their broad acceptance by the scientific and professional community (Smith, Oh, and Smith, 1989). In the early 1990s, the rapid development of computer hardware and software resulted in increased application of GIS. Notable among these developments are the improved speed, storage capacity, expert software, and relational database management systems. These advances have resulted in substantial improvement in computing economy and thus broader acceptance and application of GIS, both by the scientific community and by industry. These developments, however, have been very much influenced by the traditional methods of static development and display of maps and graphics.

The traditional GIS are static, but most Earth system processes are dynamic and require routine updating. These dynamic processes need to be incorporated in a geographic database that allows users to analyze and visualize resulting changes. The needs for analysis and display of these dynamic processes and for rapid response to policy-relevant queries by social scientists and policy makers dictate the development of dynamic GIS. In the following section I will focus on the evolving requirements of Earth scientists and some of the unique features that future GIS must have in order to be useful for global change research and policy deliberations.

8.10. Future directions in GIS development

The diversity of Earth science datasets and the complexity of methods required to store, query, and process them require database management systems far more sophisticated than currently available. Existing DBMSs are designed for commercial applications such as inventory of short records of numbers and characters. These characteristics reduce the usefulness of the existing DBMS technology for large-area (global) Earth science applications. Interdisciplinary Earth system science problems often require query and analysis of combinations of raster, vector, point, and text data together – a task that many existing data management and analysis tools cannot perform.

Earth science data records in the various forms mentioned above are not usually archived in one place; thus, network speed and access will place some limitations on the advanced applications of GIS technologies. In addition, the remote sensing component of such datasets is often of large volume. If combined with numerical simulation model results and other in situ observations, very large volumes of data may have to be transferred across the network to a remote location for analysis and display. Thus, future GIS should be capable of real-time ingestion and transfer of high-bandwidth datasets. They should also allow for presentation of this high-bandwidth, multidimensional, and multiparameter data and information in a coherent and attractive fashion beyond current static graphical techniques (see Chapter 4).

Analysis and display of dynamic Earth system processes with current GIS require multistep and lengthy computations. The user must go through every single step even if that step does not include any change. Future GIS should allow near-real-time spatial analysis and visualization of dynamic Earth system processes by bypassing the unnecessary steps that do not introduce any change in the data. In current GIS, the user must define the topology of the data and structure for it to fit the topology. For dynamic Earth system processes, this topology changes continuously. Future GIS should integrate topology into the database management tools such that it becomes transparent to the user. This in turn requires GIS to provide an automatic capability to compute topological attributes to account for the spatial change in Earth system processes. Earth scientists and policy makers also need to visualize such spatial changes as a function of time, thus requiring the addition of time as the fourth dimension for analysis and visualization in future GIS.

8.11. Conclusions

The long-term in situ and space-based observations planned as part of the U.S. Global Change Research Program, along with the interdisciplinary approach to examining the complex, dynamic, interactive processes that form the Earth system, will offer us a new perspective on how Earth functions at regional, continental and global scales during the next several decades. Knowledge gained through this initiative will benefit the public and those interested in establishing agricultural, industrial, and environmental policy by providing the sound scientific basis for such decisions. This scientific understanding is crucial to the implementation of adaptive and/or mitigating actions that may be proposed for coping with the changes that may adversely affect the habitability of Earth. New alliances should be forged among the disparate

Earth science disciplines to join in interdisciplinary efforts to address global change issues.

The planned in situ and space-based Earth observations will assure continuity in current measurements and also provide the improved spatial, spectral, and temporal resolution needed to address new and emerging scientific questions and policy decisions. These new capabilities will also provide enhanced accuracy and precision in measurements, which are needed to detect and distinguish among small changes due to natural variations and human induced activities; these capabilities are not met by current Earth observing systems. In particular, NASA's Mission to Planet Earth will provide a tremendous volume of data and information in support of these initiatives. In the next few years (i.e., from now until 1998), NASA is sponsoring several activities to provide easy access to the existing data from the Earth observing research and operational satellites under the "pathfinders" projects (Asrar and Dokken, 1993).

One of the basic challenges facing the USGCRP is how to manage the large volume of data resulting from these diverse initiatives in an effective manner and to make these data more accessible and useful to the global change research community. Remote sensing observations combined with numerical simulation model results and other in situ observations form a very large volume of data to be transferred across the network to a remote location for analysis and display. Interdisciplinary problems in Earth system science will require query, analysis, and management of this large volume of data that, rather than being archived in one place will be distributed throughout the United States and indeed around the globe.

Over the next several years, our observational capabilities, data processing and visualization capabilities, and scientific understanding of the issues identified will undoubtedly improve. This improved understanding will provide the necessary knowledge for sound policy decisions on how to preserve and protect the Earth and its environment through sound management practices. With this, we also hope to refine and improve the current data and information capabilities needed by the MTPE and EOS in support of the U.S. Global Change Research Program.

9

Research needed to improve remote sensing and GIS integration: conclusions and a look toward the future

John E. Estes and Jeffrey L. Star

9.1. Introduction

This final chapter presents the authors' opinions on the research needed to improve the integration of remote sensing and geographic information systems. The material presented represents the culmination of the authors' experience with the National Center for Geographic Information and Analysis Initiative 12, "The Integration of Remote Sensing and GIS." The chapter draws heavily on the material we prepared for the 25th International Symposium on Remote Sensing of Environment, in Graz, Austria (Estes and Star, 1993) and the final report of NCGIA Initiative-12 (Estes, Star, and Davis, 1995). One of the objectives of NCGIA initiatives is the development of an agenda that describes the research required to remove impediments to the application of GIS technology in areas under investigation. This was accomplished during the course of I-12 is presented in the final report referenced above.

The objectives of this work are to (a) broaden the dissemination of the research agenda developed during the I-12 initiative process; present an update of our views based on recent events and current thinking, (b) continue to stimulate discussion concerning the research needed to improve our fundamental understanding of the issues involved in the integrations of remote sensing and geographic information systems, (c) present conclusions and recommendations concerning needed research, and (d) present our view of a future where these two important technologies are more fully integrated. Basically, this work represents our synthesis of the research needs, issues, and topics that were raised during the NCGIA I-12 initiative process. It goes beyond the final material presented in I-12, however, providing an update based upon current activities in the area and a view of the future based upon our reading of the clouded crystal ball of current events.

In the development of the material presented in this chapter we have presented, discussed, debated, and finally prioritized the research agenda. We wish to emphasize that the material presented here is our opinion. We thank all I-12 participants for their participation in the initiative. It can be creditably argued that, at the beginning of the initiative, there quite literally was no research agenda in the area of remote sensing GIS integration. There is still no comprehensive, well-thought-out, funded research program in any federal agency that is moving research forward in this area. There are bits and pieces of the topics called out herein that are being funded, yes, but most of this funding is for other topics and the work on improving remote sensing/GIS integration is secondary to the main thrust of the funded effort. To be really effective we feel some form of coordinated effort under the direction of either an agency or coordinated by an interagency group is required. It can also be argued that the efforts of the participants in this initiative represent the first attempt to bring a semblance of order to the identification of research required in this important research area. The reader should also understand that the authors do not feel that the material presented herein is the final word on this topic. We hope that this material will continue to be discussed and debated, and that other authors or agency representatives will add and delete topics. We hope this agenda remains a living document that will help guide institutions, organizations, and individuals in the future as they fund, propose, and/or conduct research in remote sensing/GIS integration.

Following this introduction a background on remote sensing and GIS is given. The material that follows the background section describes the evolution of the I-12 research agenda. This is followed by the authors' prioritization of the research agenda and conclusions concerning our work to date in this area. Finally the authors' view of a future where remote sensing and GIS are fully integrated is presented.

9.2. Background

Remote sensing is a technology whose development has been heavily conditioned by the U.S. federal government. Geographic information systems development has been primarily influenced by commercial enterprise. There is ample evidence that for many years both of these technologies employed analog data and manual processing. Cartographic products either employed overlays, depicted multiple themes, or employed multiple displays to depict relationships between themes or changes in a given theme. The "Snow" map of the relation between wells and plague in England is an example here. Photo interpreters used aerial photography to map German forests in the late 1800s, and by the time of World War I, interpreters were analyzing thousands of aerial photographs per day, extracting tactical battlefield information.

In both remote sensing and GIS, digital processing begins to become important in the 1960s and 1970s. Indeed, even the names of these two technologies come about at virtually the same time. Ms. Evelyn Pruitt of the United States Office of Naval Research coined the term "remote sensing" in 1958. Dr. Roger Tomlinson is credited with the term "geographic information systems" in 1962. Dr. Tomlinson was working for a Canadian photogrammetric engineering firm, Spartan Aerial Services, and was involved in developing a proposal to select lands for forest plantations in East Africa. As part of this proposed effort the development of a spatially based methodology for selecting these lands was proposed. The system to be employed to implement this methodology was termed a geographic information system. Ms. Pruitt was involved in efforts of the U.S. federal government to fund research into new sensor systems that produced images of energy outside the visible portion of the electromagnetic spectrum. These systems do not take aerial photographs. The analysis of these images could not be called photo interpretation, and it was felt that a new broader term was required. And so, remote sensing of the environment began to be used to describe this emerging area of research interest.

Both civil sector remote sensing and the early use of GIS had an agriculture connection. Although the Spartan Aerial Services proposal was not funded, a fortuitous set of circumstances brought the geographic information system concept to the attention of members of the Canadian government. Specifically, to the Agricultural Rehabilitation and Development Agency that was conducting a review of Canadian agricultural lands. The Canadian GIS was developed to facilitate the analysis of Canadian Land Inventory data and identify marginal agricultural lands.

In the United States researchers at Purdue University, as well as researchers at the University of California, Berkeley and the University of Michigan, Ann Arbor, conducted research throughout the mid-to-late 1960s and into the 1970s that focused on the agricultural applications of remotely sensed data. This research was primarily funded by NASA. This was a time when the civil sector applications of optical-mechanical scanners in aircraft were beginning to be explored. These scanners produced multispectral data streams, and emphasis, primarily at Purdue University, was placed on the development of computer-assisted methods of processing these data. The work of these institutions and others, particularly in the *Earth Resources Observation Satellite (EROS)* Program of USGS, laid the foundation for the *Earth Resources Technology Satellite (ERTS)*, which was put into orbit in 1972. *ERTS* was later to receive its current name *Landsat*. The *Landsat* series of satellites are still the primary United States Earth resources observing satellites today.

Although examples of the parallel development of these two technologies can be carried too far, it is interesting to note that the first commercial GIS begin to be developed in the early 1970s time frame, the time at which significant amounts of digital remotely sensed data began to be produced. By 1976 researchers in remote sensing had begun to propose the integration of remote sensing and digital elevation data to facilitate the extraction of forest cover information. Work by Alan Strahler, then at the University of California (UCSB), Santa Barbara, and Roger Hoffer, then at Purdue, was funded by NASA. These were among the first funded efforts in the United States to integrate remote sensing into an essentially raster-based GIS context. In this early 1970s time frame, researchers at the Jet Propulsion Laboratory in Pasadena, California; NASA's Earth Resources Laboratory at Slidell, Louisiana; Goddard Space Flight Center at Greenbelt, Maryland; and Ames Research Center at Moffett Field, California were also pursuing the beginnings of raster-based remote sensing/GIS integration. The "genie was out of the bottle" so to speak, but proliferation of research here was, and continues to be, limited by a wide variety of factors and circumstances. Removing and/or reducing these impediments to the fuller integration of remote sensing and GIS is the driving force behind the research agenda presented herein.

9.3. Evolution of a research agenda

Research agendas are, at a minimum, difficult to develop, worse to prioritize, controversial, temporally fragile, and subject to the very real biases of their authors. We know at this time of no organized, prioritized research agenda that is being employed by any organization to direct fundamental research in either remote sensing or geographic information systems. Thus, trying to develop a prioritized research agenda directed at the integration of remote sensing and GIS is likely somewhere between a square function and an order of magnitude more difficult than the development of a prioritized research agenda for either topic alone. Yet, that was the task that the co-leaders of I-12 set out to accomplish. The material that follows is very much the authors thoughts and carefully considered opinions on this subject at this time and is only a step in a process; it is not an end in and of itself.

Table 9.1. *Preliminary prioritization*
based on frequency of mention by I-12
expert meeting subgroup

# of Group listing	Issues
5	Standards
4	Lineage Tracking Raster/Vector Visualization
3	Accuracy/Error Education/Training Scale Test Data Sets
2	IGIS Taxonomy Multistage Sampling Spatial Data Catalogs Spatial Statistics

The I-12 initiative process began with the first meeting of the small steering committee in Denver, Colorado and was further developed at a meeting at Stennis Space Center, Bay St. Louis, Mississippi (see Chapter 1 for further details). These meetings of the steering committee of I-12 set the stage for the specialist meeting held at the USGS EROS Data Center (EDC) in the fall of 1990. The results of the discussions that went on at this meeting were presented in a special issue of *Photogrammetric Engineering and Remote Sensing* in the spring of 1991. As a preliminary step in the overall process of developing a prioritized research agenda, one of the authors reviewed the results of I-12 activities accomplished through September 1991 (Estes, 1992). As part of this analysis a matrix of common research themes was developed. The matrix was developed by a review of the discussions surrounding each of the five research areas addressed at the I-12 specialist meeting. The objective of this review was to identify where research topics were called out as important by I-12 participants in more than one of the five subgroups (Institutional Issues, Data Structures and Access, Data Processing Flows/Models, Error Sources and Analysis, and Future Computing Environments). We realize that this is an imperfect way to assess the significance of given research areas and topics. Nevertheless, the listing of research needs in this fashion does represent one way to portray the relative thinking of the remote sensing/GIS community concerning significant research issues that require attention.

The results of this exercise can be seen in Table 9.1, where the research themes gleaned from an analysis of the individual I-12 specialist meeting subgroup findings are presented in descending order of priority. Themes are ranked by number of groups identifying each topic area. By ordering the research themes in this fashion, we can develop a rough-cut prioritization of areas where specialist meeting participants felt that research was required.

This ordering also serves to highlight generic areas with respect to the subdivisions adopted for the conduct of the expert's meeting. *Standards*, for example, although not particularly a major topic of discussion in any one subgroup, were nevertheless recognized as a priority topic by all groups. Participants in all five subgroups believed that more emphasis on the development of data, hardware, and software standards was required. They believed that more attention to standards development was particularly needed in the GIS area and in the whole area of spatial data analysis as well.

In the second tier of prioritization, we find those categories listed by four subgroups. In each case it was the Institutional Issues subgroup that did not mention the topic. *Lineage Tracking, Raster-to-Vector Conversion*, and *Visualization* were considered priority items by all of the remaining subgroups. The authors do not consider this unusual given the subgroup themes. Each of these categories is clearly important to the subgroups mentioning them (Data Structures and Data Access, Data Processing Flows Error Sources and Analysis, and Future Computing Environments).

As we move to the next level, the reasons for listing, or not listing, become somewhat more complex. *Accuracy and Error* was not listed by the Future Computing Environments subgroup. The authors do not feel that this is particularly unusual given that this subgroup was primarily focused on issues related to system development. *Education/Training* was not listed by either the Data Structures and Access or the Processing and Flows subgroups. This is likely because these two groups were largely concerned with techniques, methods, and science development, whereas the other three groups which did list *Education/Training* were possibly more concerned about improving our overall quality of understanding of the technologies and the synergism associated with their integration. However, it is still somewhat difficult to understand why *Education/Training* was not mentioned, in particular, by the Data Processing Flows subgroup. Agreement on the topic of *Scale*, in contrast, is more straightforward. The reader can appreciate to some extent why the Institutional Issues and Future Computing Environments subgroups might not have listed Scale as an issue of central concern at this time, whereas the Data Structures and Access, Data Processing Flows, and Error Sources and Analysis groups would be very concerned about issues of scale. Yet, issues of scale are currently very important in supercomputing, and institutions are trying to find better ways of dealing with a variety of scales of data for many types of analysis. Finally, the reader can see some of the same rationale in the listing of *Test Data Sets* by the same three subgroups that listed scale. The need for the development of a set of data-rich areas where the accuracy of existing datasets has been as thoroughly verified as practical is critical. These areas could be used as test sites for a wide variety of experiments designed to improve our understanding of the issues that impact remote sensing/GIS integration. Test sites have been developed on a program-by-program or project-by-project basis in the past by a number of U.S. and foreign governmental agencies. The authors believe that there is a need for greater cooperation and coordinated use of the test sites that are either already developed or those that are in the process of being established.

When we examine the topics listed by two subgroups, the rationale for listing is fairly straightforward. *IGIS Taxonomy, Multistage Sampling*, and *Spatial Statistics* were each considered significant by the Processing Flows subgroup and the Error Sources and Analysis subgroup. The relevance of each of these research themes to these subgroups should be readily apparent to the reader. *Spatial Data Catalogs* are significant as a means to improving our ability to locate and acquire data. The need for additional research in this area was listed by the Institutional Issues subgroup and the Data Structures and Access subgroup. The authors believe that, here again, the reasons for these two subgroups listing this topic are intuitively obvious, given their particular focus.

With respect to the prioritization depicted in Table 9.1, it can be argued that the relative listings were to some extent preordained by the I-12 executive committee. This preordination basically occurred when the five subgroups were chosen at the second executive committee meeting. Yet, it can equally be argued that these subgroups were then put forward to the specialist meeting as a whole and that these five subgroups were validated by I-12 specialist meeting participants. It was the specialists specifically chosen for their expertise in remote sensing/GIS integration in each subgroup that are responsible for the topics chosen and their relative rankings. As such, we must seriously consider that work is required in each of these areas and that the prioritization of topic areas where research is required, seen in Table 9.1, is valid in and of itself.

It is difficult, however, for the authors to fully justify this prioritization if we examine research needs in specific applications areas where the use of remotely sensed data would be employed to create or update key data layers in a GIS being employed as a decision support system. In areas like global change or biodiversity research, just as with location analysis or facility siting, the ability to more effectively and efficiently integrate remotely sensed data into GIS-based analyses is critical. We need to improve the accuracy of information derived from remotely sensed data using processing techniques that take advantage of a priori information, scene-based models, and/or site scientific data. We need to be able to directly rectify, ingest, and georegister these data with spatial and nonspatial data within a GIS. We need to be able to more efficiently query very large spatial data bases. We also need dynamic n-dimensional visualization of integrated remote sensing and GIS data layers in output products. If we consider these as examples of essential areas of research need, we could be led to place higher priority on understanding key research issues in topics such as classification and feature extraction, error propagation and accuracy assessment, multistage sampling, and spatial statistics, to name but a few areas. Here also, data structure and format conversion (raster/vector) questions and issues of scaling gain in significance.

Basically, what we are saying is that any prioritized agenda can be seen as being user or constituency driven. In the case of the prioritization seen in Table 9.1, no end use or user community was specified except those broad communities agreed upon by each of the individual subgroups. In addition, it is important to take into consideration that participants at the specialist meeting were not asked if research was being conducted in an area nor, if research was being conducted, was it being adequately staffed and/or funded. In the opinion of the authors, most subgroups identified areas where key impediments did, in the considered opinion of participants, indeed remain. And just that! There was little or no within-group prioritization. Indeed, there was little time for such prioritization. There was also no attempt at across-group prioritization at the specialist meeting. Time was a critical constraint, and to our knowledge, only the coleaders of I-12 have made any overall attempts at prioritization in this area.

The authors of this chapter believe that both the real and the potential levels of effort and funding required to support research should be considerations in the development of a prioritized agenda in any area. Is an area of research already being worked? By whom is it being worked? Are the efforts adequate? Are more resources (personnel, equipment, funding) required? From a purely pragmatic perspective, questions regarding types and levels of activity with respect to a given research area need to be examined, evaluated, and taken into consideration in the development of a research agenda. To some extent, simply the listing of a topic may seem to imply that the specialists considered the research activity currently underway in the area inadequate. Is this really true? Was this the case with respect to the participants at the I-12 initiative meeting? Perhaps; perhaps not.

We feel the need to address the question of the adequacy of ongoing current and planned activity in a given research area is particularly important when we remember that remote sensing is a technology heavily influenced by developments in the federal establishment,whereas GIS technology has been largely driven by private industry responding to market pressures. Whoever decides to fund or conduct research in this area must examine what research has been and is being done. This should be done by anyone before deciding what research is required, should be initiated, and which areas should receive priority attention. The authors realize that these seem like trivial and self-evident statements. We make the point of saying them here because we understand the difficulty of the assessment that is called for in trying to survey activities in fields as dynamic as remote sensing and GIS, let alone the integration of the two. Trying to survey activities in academia, government agencies, nongovernmental agencies, and private industry in both the civil defense and intelligence communities around the globe is truly no easy task.

Thus, the authors feel that there is more than one path that can be taken in establishing priority areas requiring research. We also feel that many arguments can be made concerning the adequacy of research and resources currently being directed toward any given area. Yet, the development of a prioritized research agenda in the area of the integration of remote sensing and GIS is what the authors set out to do in this chapter; and what, with the background gained from the initiative process and our years of work in the field, we will do.

The prioritized research agenda that follows is our considered opinion, our best judgment at this time, of where considerable resources need to be placed now! By resources in this context, we do not only mean dollars. We mean adequate dollars, yes, but that is only part of the need. We also mean personnel, facilities, hardware, and software resources as well. We feel that if such resources are made available, key impediments to the fuller integration of remote sensing and GIS can be overcome and a wide variety of basic and applied research issues can be more effectively addressed. We also feel that the potential for the operational application of these technologies will be significantly improved if research in the areas suggested bears fruit, as we believe it should.

Remember as you read the material presented below the admonitions given earlier in this paper: that the material presented here is part of what must be an ongoing process and that as we improve our understanding of the issues we must continue to update this material, reordering our priorities as appropriate.

Table 9.2. *The authors' prioritized research agenda*

---

1. Science and Technology Advancement
    A. Advanced Feature Extraction
    B. Spatial Analysis and Modeling
    C. Visualization
    D. Lineage or Heritage Tracking
2. Improved Understanding
    A. Education and Training
    B. Data Format and Structure Conversion
    C. Spatial Information Management
    D. Error and Accuracy
    F. Scale
    G. Time
    H. Heterogeneous Computing Environments/Interaction
3. Infrastructure Development
    A. Standards
    B. Spatial Data Catalog
    C. Test Datasets

9.4. A prioritized agenda for remote sensing and GIS integration

The authors take as a baseline that:

(1) A number of research topics were identified by I-12 participants that are currently being aggressively pursued by government, nongovernmental organizations, industry, and/or academia.

(2) I-12 subgroup participants were primarily focused on subgroup priorities.

(3) This allowed certain important issues to "fall through the cracks."

(4) There was little discussion among specialist meeting participants of overarching or critical interest priority issues.

(5) Such discussions were primarily conducted after the specialist meetings at symposia, and conferences and in formal and informal gatherings of interested parties.

Based upon this somewhat shaky foundation then, we begin our "current/final" prioritization with the premise that we will have in our highest priority category, Science and Technology Advancement, those topic areas where it is our opinion that the most significant near-term scientific and technological advances in the integration of remote sensing and GIS can be achieved (see Table 9.2). Basically, these are research areas where we believe insufficient resources are being expended at this time to adequately address the nature of the fundamental issues requiring examination.

In our second layer or tier of priority topics, Improved Understanding, we include areas that, despite their importance, we believe will take longer for researchers to make significant progress. Included here too are several areas where research is already being accomplished. But, in these areas more resources, in terms of both personnel and funding, are required to adequately address the issues involved. We wish to make it very clear here that we believe significant additional funding is required over and above that which we know are currently being directed at the topic areas listed in ranks one and two of Table 9.2. It is our hope that federal or international funding agencies, nongovernmental organizations, or private industry will take notice and begin to provide the level of resources necessary to effectively conduct the research needed in each of these areas.

In the third tier, Infrastructure Development, we list topics that I-12 participants believed to be important but that we feel are currently being addressed with sufficient resources at this time. This is not to imply that the work required in these areas is any less important. Nor does this imply that levels of resources currently being directed at these areas could be either reduced or redirected to topic areas in tiers 1 or 2. What we mean here is that this work must continue, but we cannot recommend significant increases in funding for these specific areas of remote sensing/GIS integration related research needs at this time.

It is important that the reader should not be left with the impression that we consider the funding for research in these important areas a zero sum effort by any stretch of the imagination. We do not. And more importantly, we feel it is not. There are serious research areas and topics that must be addressed at all levels, and it is up to all of us, the reader included, to make this case. We believe remote sensing and GIS are two very powerful technologies whose full potential is far from being realized. These technologies offer the best hope that we have for the wise use of data for both conservation and development – technologies that are critical if we ever hope to achieve sustainable economic development. The work here will require funding for personnel, hardware, and software development. It will require management attention. What is critical, however, is that the full potential of these important technologies will not be reached unless we are successful in expanding current fundamental, applied, and pre-operational research activities in this area.

Viewed another way, we have chosen as our highest priority recommending expanded research for those areas that we believe will advance the scientific, technical, and applications potential of integrated geographic information systems (IGIS) in the most effective and efficient fashion. For tier 1 topics, we believe that significant progress in these areas is possible in a two-to four-year time frame given increased attention, including funding opportunities from agencies, other organizations, and/or directed research funding from private industry.

In our second tier, Improved Understanding, we have placed topic areas that, though important, we feel:

- are generic, inherently difficult issues that will significantly test our ability to demonstrate acceptable results (e.g., *Accuracy and Error, Scale*, and *Time*);
- represent current work, that although promising in some areas, will be more difficult and time consuming than some think at this time (e.g., *Heterogeneous Systems Environments*); and
- are longer range issues that do not lead themselves to "solution," but must be continuously watched as the technologies and the science evolve (e.g., *Education and Training* and *Spatial Information Management*).

Significant advances in these areas will require sustained commitment on the part of both the funding and the research community to research. It is our opinion that the issues that need to be addressed here are of a nature that a five-to fifteen-year time frame will be required before research and development activities in these begin to show significant benefits with respect to operational IGIS applications.

In our third category, Infrastructure Development, we have grouped three topic areas: *Standards, Spatial Data Catalogs*; and *Test Data Sets*. Developments in this area of activity could impact all segments of the remote sensing and GIS communities from system developers to research and applications users. These are topics that we believe are important, but that are being addressed within the community at what we consider to be an acceptable pace at this time. We say this even though many of us wish standards efforts, in particular, would move at a much faster pace. The authors are not the only ones who feel this way. Other groups are also calling for a speed up in the standardization and harmonization of a variety of issues impacting the integration of remotely sensed data into GIS processing and analysis flows.

It is important to note that just because a topic is listed in this Infrastructure Development category does not mean that it is any less deserving of support than topics included in tiers 1 (Science and Technology Advancement) or 2 (Improved Understanding). It does not. Although this may at first seem somewhat confusing within the context of this prioritization, it is really quite straight forward. Topics related to Infrastructure Development are being worked by NASA, USGS, the U.S. Environmental Protection Agency (EPA), DOD, and a variety of institutional, industrial, and academic participants. This work must continue. This work must be funded. But, more significantly, this work must be better coordinated at a variety of levels. Progress and the result of this work must be made more widely known and be made more accessible to workers in both the remote sensing and GIS communities. Methods and efforts to effect this improved coordination and dissemination must be pursued with vigor. We must do a better job of involving more of the whole of the global remote sensing and GIS community in these efforts where appropriate.

The research areas shown in Table 9.2 are listed in priority order. This is "our call" based on our understanding at this time. We believe the area deserving the highest priority at this time is *Advanced Feature Extraction*. This was a topic that received less attention than the authors feel it deserved at the I-12 specialist meeting. The topic does not fit neatly into any of the I-12 sub groups. Its best fit was in the area of processing flows/models. Yet, there was little focused discussion on this specific topic at the I-12 specialist meeting. Subsequent discussions and the needs of programs such as the global Advanced Very High Resolution Radiometer (AVHRR) 1-km land cover project, the United Nations Development Program (UNDP)/United Nations Environment Program (UNEP) core dataset efforts, GAP analysis, National Wetlands Inventory, and E-MAP Land Characterization Programs make this a very important area for research focus. Indeed, we do not feel that this area has been given the attention it deserves for more than a decade. There are pockets of research activity in this area in the community and some very good work is being done. There are and have been for years, efforts directed at the extraction of "point targets" by elements of the defense and intelligence establishments. Yet, we are aware of no major funding agency that has a truly focused program of funded research in this area. There definitely should be at least one major program effort here, and likely more, directed at advanced feature extraction across the full range of features of interest to the community.

The background for this lack of research goes far beyond the scope of this chapter. We believe such a discussion would make an interesting thesis or, given the correct circumstances and access to material, dissertation topic. Suffice to say that the remote sensing/GIS community, largely due to lack of funding opportunities, has not pursued research in this area. Some major global change research efforts focused on tropical deforestation and biodiversity vegetation classification projects initially employed manual image analysis techniques from single-band satellite imagery; others have just moved to simple unsupervised classification. These efforts are symptomatic of the problem. When we are still using unsupervised clustering and band rationing of AVHRR data for regional scale, change detection, and product generation we have definitely missed something along the way. We must stop and rethink our activities in this area now. We must begin to explore the full potential of integrated GIS/remote sensing analysis for feature and information extraction from remotely sensed data.

Our second area in terms of priority is the broad topic of *Spatial Analysis and Modeling*. This is a critical area of research need. It could be very persuasively argued that this area should deserve the highest priority. We have chosen not to give this area that rank largely because we feel that advances in tier 1 are required before we can fully realize major advances in this area. This does not mean that significant advances in the integration of spatial analytical techniques and modeling cannot be accomplished now. They most certainly can. The evolution of object-oriented database technologies is leading toward the development of new integrated data models for managing both vector and raster data. Research activities should be pursued and funding opportunities must be expanded dramatically, in both *Advanced Feature Extraction* and in *Spatial Analysis and Modeling*, if we hope to apply integrated remote sensing and GIS technologies to satisfy the information needs of the wide range of potential users of these data. Users all the way from the global change research scientists at UCSB to the city or county planner in Buffalo, New York, to the forester in Maine, to the farmer in the field in South Dakota, to the lawmaker in Washington, D.C. or the regional planner in Niamey, Niger, or the project engineer in Chaing Mai, Thailand, for that matter, all have a need to get more accurate, up-to-date information concerning components of our environmental envelope. Work in both these areas can significantly benefit all these and other types of users as well.

Federal agencies are spending large sums to implement GIS within their organizational structures. These agencies expect that the systems they purchase will satisfy their requirements and that vendors will continue to advance the applications of these systems. In reality, what typically happens is that as agency personnel begin to work with these systems they discover that the system cannot perform some of the operations which they thought they would. This is particularly true in organizations where various types of modeling are important. This has had, to some extent, a negative impact on the field and has, we believe, negatively impacted extramural funding by these agencies. This lack of support for modeling operations is a problem that must be faced by both the remote sensing and GIS communities today. To date, vendors have largely concentrated their development activities on those areas where they feel they will achieve the greatest economic return. Today, most GIS users essential employ their GIS as sophisticated electronic file cabinets and use GIS functionality largely for relatively simple overlay operations. Most GIS vendors are generally adverse to examining potentially high-risk research areas,

even if such research may have high, long-term pay off. Vendors have thus
focused on those functions, such as overlay and search, that allow users to
employ GIS systems as electronic file cabinets for their spatial data. Although,
there is some evidence that this situation may be changing we believe things
are not yet changing fast enough to help in a variety of high-profile research
areas.

The third topic in our Science and Technology Advancement tier is *Visualization*. Despite the great deal of activity currently going on in this area, more
still needs to be done. Simulations of the impacts of changing land cover patterns on convective cells near airports, changing urban patterns over decades,
and stream turbulence are but a few examples of areas where research is being
conducted on more effective ways to communicate IGIS output to users. Such
users need to be able to more effectively view their analyses as they progress
in real time or, at a minimum, "real-enough" time. We must be able to convey
attributes of data more effectively. We need research directed specifically at
optimal methods for the presentation of data to both public and private decision makers in a variety of areas and at all levels. We must move from 2D
and 2.5D graphic visualization to more realistic 3D and 4D graphic visualization. Such developments would add significantly to our understanding of a
wide variety of complex and interconnected socioeconomic, biophysical, and
geochemical processes occurring around us.

The final area in the first tier is the area of *Lineage* or *Heritage Tracking*.
This is an important area from both a science applications and a technology
perspective. We need to track the history of the datasets we are using in our
analyses and we also need to maintain a record of what we do to those datasets
as we move through a processing and analysis flow. Decision makers, analysts,
policy formulators, and the legal community all need to be made more aware
of the history of the data employed in a given task and how the processing of
that data might impact its efficacy for a variety of applications. We believe that
there have been significant advances in this area in the past several years and
that an investment in this area now could bring to the community significant
benefits in very short order.

The authors realize that in this prioritization there are several areas where
it might be argued that required research is properly the focus of industry.
Lineage/Heritage Tracking is certainly one of these areas. Here too we could
include the tier 2 (Improved Understanding) topics of *Spatial Information
Management* and *Heterogeneous Computing Environments* as deserving of
attention by primarily the commercial community. These three topics are
certainly deserving of attention and important to a wide user community.

However, we believe that significant improvements in the state of the art in these areas would not be most effectively and efficiently accomplished in either the academic community or governmental agencies. We believe that private industry should assume a lead role in these areas.

This conclusion is just one more example of the authors' bias. It is our opinion that the private sector is probably more aware of the requirements for these kinds of operational capabilities and is potentially more capable of moving rapidly with prototyping and joint development with users. That does not mean that academics or government researchers cannot, or should not, participate in work here nor that the legitimate need of government agencies for improved capabilities in these areas cannot be used as a justification for the conduct of research on these topics. It can, and indeed is, being used as such. What we are saying is that industry needs to be a major player/partner in the work that must be accomplished in these three topic areas in particular.

For the purposes of this discussion we have chosen the preceding topic areas as being focused on Science and Technology Advancement. Again, we feel that research and development can have higher near-term payoffs for the broad user community than are likely to occur in those areas that follow. We use the term "Improved Understanding" to describe what we believe are the next highest priority research and development activities. At other times we have used the word "Wisdom" to describe this tier of research topics. We believe this second tier of topics will require more sustained, focused research and development (five to fifteen years) for solutions to be optimally achieved. Indeed, it is even difficult to say that optimal solutions will be achieved here, since what might be optimal for one class of user may not be for another user.

It is not surprising that researchers from an academic institution would put *Education and Training* high on any prioritized list. As the reader will soon realize there are several topic areas in this second tier that could either go into our first or third level of prioritization. It might be argued by some that *Education and Training* is more an infrastructure issue and should be included in our text tier of research priorities. In one respect we agree. Education and training are crucial to a solid infrastructure. Yet, one of the criteria for inclusion in the next tier of topics is that activities in these areas are being adequately addressed and funded at present. In the area of training and education this is certainly not the case. Although there are a number of efforts currently underway to improve education and training in remote sensing or GIS, the authors are aware of only one that focuses on the integration of remote sensing and GIS.

This effort, which is still somewhat in the formative stages, does not in our opinion currently have sufficient funding or a level of community participation to carry the project to successful completion. As an outgrowth of NCGIA I-12, a group of academic and industry personnel are seeking the resources required for such an effort. A curriculum development project in remote sensing and GIS deserves funding agency support at levels from kindergarten to graduate school. Development of curriculum materials in these areas can provide students at all levels with graphic aids and text to enhance learning. There is an old Chinese saying: "If your horizon is one year, plant rice. If your horizon is ten years, plant trees. If your horizon is one hundred years, educate your children." Mankind is here, hopefully, for the long term. We must do a better job of educating people to the true potential of these combined technologies.

In the area of *Data Format and Structure Conversion*, in particular, we need to be able to convert data from one format into another and to move between raster and vector data more effectively. While some may argue that this area represents a "done deal," the authors do not. In particular, although there is certainly software with these functions, in our view there is not an understanding of the effects of these functions on the quality of the resulting derived data layers. Notwithstanding the numerous systems with a "format conversion button," there is still insufficient understanding of how these functions affect the decisions we reach with these software tools. We do, however, feel that significant advances can be made here, and we encourage funding agencies and individuals or groups with innovative approaches to problems in this area to redouble their efforts.

Spatial Information Management is a very important area for research and development. Like *Lineage/Heritage Tracking*, this is an area where industry should be heavily involved. As the amount of digital data in spatial format grows and repositories of this type of data spring up around the nation and the world, the need for improved management concerns and tools increases. An indication of the importance that is being placed on research and development in this area is the NASA/NSF/Advanced Research Projects Agency (ARPA) funded digital libraries projects. These projects, funded in the millions of dollars, are addressing important data storage, management, communication, and access issues. Project Alexandria, an effort that includes NCGIA consortium members, is the only one of these projects that is specifically focused on spatial data. The research and development efforts associated with the Earth Observing System Data and Information System (EOSDIS) are also critically important here as well. So, what we see is that work is being done here in a number of areas, but industry does not seem to be taking the lead. The question is why? Quite frankly, other than the reasons suggested above, the authors have no idea.

An area of particular interest and importance, which we feel is not receiving sufficient attention, is the whole area of spatial database management systems (SDBMSs). Traditional database management systems largely evolved from COBOL account programs and FORTRAN/UNIX-based data management systems. These systems do provide users with powerful statistical and modeling capabilities. Analytical constructs, however, are limited to aggregations of locational and spatial relationships. At present, fundamental spatial data relationships such as connectivity, proximity, and contiguity are difficult to explore using traditional database management systems. Lack of capabilities here limits the utility of these systems in many research and decision-support studies in areas such as environmental assessments, human-social interactions, health care studies, and many others.

Spatial data management requires particular attention to the locational attributes required for spatial analysis. Although spatial data management requirements vary from exacting topological data structures to labeled and linked spatial identifiers, the systems management overhead within an SDBMS is not trivial and cannot be easily dismissed. The tremendous growth in business and marketing use of the spatial data management tools for economic modeling also requires attention to the liability associated with data usage. Environmental and health risk assessments, which are dependent on spatial data management capabilities, reinforce the critical need for research in this area.

Current knowledge of, and the teaching environment for, spatial information management is extremely limited. These limitations have imposed serious restrictions on a variety of spatial data management issues, which must be addressed in the classroom. Some of these include: (1) data structures; (2) object-oriented paradigms for data, programming, and systems; (3) database exchange formats and standards; (4) temporal updates and maintenance of spatial databases; and (5) raster/vector data integration. Significant research remains in the domain of SDBMSs, as attested to by the serious obstacle and failures of many large projects or programs dealing with this particularly important research area.

Error and Accuracy is certainly an area where a great deal of effort is needed. This topic was originally listed as a tier 1 topic and has been moved back and forth between tier 1 (Science and Technology Advancement) and tier 2 (Improved Understanding) several times. There is no doubt that error/accuracy is a very important area and deserves considerable attention. Our reasons for listing this area at this level, at this time, is that we are not entirely sure that significant progress can be made in this area in a one- to four- year time frame. Indeed, the same thing can be said with respect to timing for each of the next two topics, *Scale* and *Time*.

Each of these three areas – *Error and Accuracy, Scale*, and *Time* – are fundamental attributes of spatial data, and while it can be argued that several of the other topics in this priority listing also include issues related to important attributes of spatial data, we would argue that these three are really the most generic. Improved understanding of the nature and magnitude of these areas is a very real key to improvements in understanding that go far beyond the bounds of even this broad initiative. Long-range, thorough research is certainly needed in each of these areas. A problem here is that research of the type needed is not currently high on the priority lists of major funding agencies nor of private enterprise. The reasons for this are primarily related to the near-term return motives of industry; the charters, roles, goals, and missions of individual agencies; and the level of funding resources currently available to these groups.

Few agencies feel that their missions include the necessity to pursue research in these areas; even those that do such research realize that there are other issues of priority importance where there are significant shorter term payoffs. This is somewhat of an analogous situation to industries' focus on the bottom line. If it can't show quick profit, why pursue it? Basically, both industries and agencies know enough to make sure that the work they do shows significant returns within their annual funding cycle. Yet, it is precisely this type of research that the authors strongly feel federal agencies should be funding. The authors are aware of some federal agencies whose budgets are so small, and the demands on these scarce resources so great, that efforts to mount research efforts in these areas are not possible. This is basically the case with NSF's Geography and Regional Science Program. NSF is the one agency where, in our opinion, research in error/accuracy, space, and time in both remote sensing and GIS truly fits. Yet, to even begin a research program in these critical areas, the funding level in this program would need to be increased significantly – at least by an order of magnitude. The authors strongly recommend that this happen. We believe that there are important issues to be addressed here and that there is a cadre of researchers who, if sufficient resources were available, are capable of making progress. We believe that research funding in these areas would improve the overall leadership of this nation in spatial analysis and go a long way toward helping the United States maintain its technological leadership in this area.

What is happening in the research being conducted in these areas (error/ accuracy, scale, and time in remote sensing and GIS) today, in our opinion, is that we are largely picking around the edges. Researchers working on other topics are learning how problems in dealing with error/accuracy, scale, and time impact their particular basic science or applications issue. Yet, since these issues are not the focus of their work they do not receive the directed attention they deserve. Perhaps, this is as it should be. Could it be that the problems here are so complex, and the benefits to be derived from an improved understanding so small, that major research efforts in these areas is not warranted?

After all, we have been, are, and will continue to make decisions based upon data whose error and accuracy attributes we do not fully understand. We cannot deny that data and information generated at one scale are extrapolated to other scales and that at times this is done without sufficient concern for the impacts that sample design, spatial autocorrelation, or multiple collinearities might have on the results at the new scale. We all know or should know of instances where data acquired at one point in time have been used to verify data collected at another point in time. These are only examples from a long list of ways in which we as researchers and applications users have both used and abused error/accuracy, scale, and time. There are good and valid reasons for doing so, but we should make it more clear to all that we understand what we are doing and why.

Particularly important is the why, not only because the understanding of the implications and impacts of these topics are not fully known, but equally because a matter that addressing these issues within the context of a given project budget can be prohibitively expensive. As was recently pointed out in several meetings on similar, but unrelated, topics attended by the lead author, it can cost more to verify the accuracy of a map than to produce it in the first place. The question here is: Can we continue to live with the current situation?

One potential approach to the major issues in these areas is that we recognize that error/accuracy, space, time, and indeed other fundamental issues in the spatial data analysis area are topics of very serious concern – topics where long-term fundamental research efforts should be encouraged. These are areas where major long-term, well-thought-out proposals from individuals, groups, or institutions need to be solicited. Indeed, if we can have Long Term Ecological Research Sites (LTERs), a long-term, well-funded center for atmospheric research such as the National Center for Atmospheric Research (NCAR), or the multiplicity of national laboratories for research in various aspects of physics (e.g., Los Alamos, Oak Ridge, Fermilab, etc.), why shouldn't we have similar well-funded center or agency-based efforts on spatial analysis

topics where an improved understanding can lead to better decision making
and an improvement in our ability to manage our planetary resources and
move toward truly sustainable economic development for the benefit of all
peoples of the Earth? More could and should be said and done here, but again
it would carry us even further beyond the scope of this paper.

The final issue in tier 2 is *Heterogeneous Computing Environments*, an
area, again, where the authors believe industry must take a lead. We need to
be able to seamlessly work between a variety of systems if we are to achieve
the full potential of distributed IGIS analysis. Platforms should all be able to
communicate with each other. Software systems should be fully compatible.
A user should be able to easily input, for example, an IDRISI file into ARC-
INFO or ERDAS. Does this require standardization and harmonization? What
will it really take to make it easier for all users on any type of system to
efficiently communicate and exchange data as a first step and to interactively
work together as a long-range goal?

Our final tier of topics is labeled Infrastructure Development. These are
topics that we feel are currently being worked with sufficient resources. Included here are three topics: (1) *Standards*, (2) *Spatial Data Catalogs*, and
(3) *Test Data Sets*. Research issues here are important and, with the very
arguable exception of the area of *Test Data Sets*, are areas in which federal
agencies, industry, and the academic community are moving forward at this
time. What must be said here is that we do not agree with all the activities
in this area or with the overall pace of these activities. We strongly believe,
however, that the initiation of any major new efforts in these areas, at this time,
would be counterproductive. That is, there is no way that any group should
undertake to duplicate the standards activities that are being carried forward
under the auspices of the Federal Geographic Data Committee (FGDC) or the
Interagency Working Group (IWG) on Data and Computation. These groups
already interface with a wide segment of the community and have established
review and oversight mechanisms already in place. Their work can and indeed
must continue for the good of us all.

An observation that must be made, however, is that there are many who
wish that the efforts of these organizations would move faster. Let us hope that
it does not take ten years to implement the raster profile of the Federal Spatial
Data Transfer Standard (SDTS). In this area it may be better to implement
and evolve a less than "perfect" standard rather than to continue to seek the
"holy grail" of perfection and delay implementation. Standards purists may
blanch at this, but scientific and applications research oriented investigators
and operational users need standards now! We need standards that facilitate

access to data, that improve our ability to move data between computing environments, and that do not put an undue overhead on the research community. There are many issues here, and we believe that at some appropriate time the broader industry, research, and applications communities must become partners in the standards development efforts currently underway. This must be done to ensure that users are appropriately served by standards, not unduly constrained by them.

Spatial Data Catalogs are being developed by a wide variety of organizations today. The important, bottom-line issue here is improved access to data, spatially referenced or otherwise. At meetings the authors have attended all around the world, access to and improved availability of data is a key theme repeated over and over again. The National Aeronautics and Space Administration, the United States Geological Survey, the United States Environmental Protection Agency, the United Nations Environment Program, the United Nations Development Program, the International Geosphere/Biosphere Program, the Nature Conservancy, the United States Research Libraries Association, and the Digital Libraries Research funded jointly by NASA and NSF are but a few examples of organizations that have or are focusing on this problem. Nevertheless, obviously more needs to be done more rapidly. At the core of most current efforts is the development of metadata. Metadata has been defined as data about data. Issues of access to metadata and metadata formats and structures overlap to some extent the research and development needs discussed in the standards area. Interestingly, with respect to the discussion on standards, metadata standards are an important issue that is actively being worked at this time.

Overarching this whole metadata area, however, are questions of networking and online access to all types of data. Advances here will require more coordination among major data depositories and providers and stronger efforts directed at catalog interoperability. Here too, catalog interoperability overlaps needs in the area of *Heterogeneous Computing Environments* and serves as another demonstration of the complex interconnectivity of all the areas addressed in this initiative.

The authors' final area of priority is the development and wider use of *Test Data Sets*. We feel such data sets are required to fully evaluate procedures developed in a number of other topic areas (e.g., advanced feature extraction or processing flows/models). There are currently only a very limited number of well-documented, verified datasets in differing environments. These existing verified/validated datasets need to be better utilized in exploratory research and development to test ideas, explore the potential of new concepts, and evaluate alternative methodologies. Test datasets could improve our fundamental understanding in a wide variety of areas. Areas such as feature extraction, system corrections, and issues of accuracy and error could be more readily evaluated and more confidence might be placed in the results of research. Without documented study sites we will always be subject to the question, "how accurate is your ground truth?" Well-documented test datasets may not fully answer this question, but we believe they can go a long way toward helping us improve our understanding of remote sensing and GIS.

A number of well-documented and verified datasets already exist or are under development. Examples here include the EPA Chesapeake Bay dataset, the NASA Ames Research Center's Oregon Transect, the First International Satellite Land Surface Climatology Project (ISLSCP) Field Experiment (FIFE) test site in the Konza Prairie of Kansas, and UCSB's Goleta and Lompoc USGS 7.5-minute Quadrangle datasets, to name a few. These sets of data cover areas of varying sizes, many of whose attributes are known and for which field verification data exist. There are more out there. Currently, both NASA and IGBP are developing a suite of test sites for *Landsat* data in areas around the world. This activity is part of the NASA *Landsat* Pathfinder effort. The test sites are being coordinated by researchers at the Desert Research Institute in Reno, Nevada.

What we believe is needed is some central focus, or clearinghouse if you will, for letting the community know that these datasets exist. Equally as important, however, is information on how users can gain access to the datasets. This whole area of storage, access, and improved information dissemination deserves considerable attention and should be taken on as a high-priority task by some institution. This task is not small. It goes well beyond the development of a home page on the internet. Continuity is critical, and to be of optimum value the information must be kept current. This need for continuity of operation argues that this is essentially a service function that should be taken on by an appropriate federal agency. The authors would recommend the USGS EROS Data Center (EDC) as a logical place for the land component of such an activity. We believe that the EDC, as with the National Satellite Land Remote Sensing Data Archive (NSLRSDA), is the logical place for information concerning these test sites and for the actual datasets themselves to reside. Such data could, like future *Landsat* and/or EOS data, be made available to users at the marginal cost of reproduction. We believe the ability of a researcher to test their algorithms on data of known quality from a wide variety of environments around the globe would be a major boost for these technologies. It would help speed the development of software and applications that could have significant impacts on our ability to measure, map, monitor, and model our global system.

9.5. Conclusions

The material presented here has significant implications for all of us who are interested in the improved integration of remote sensing and GIS. It deserves careful attention. The reader should understand that governments of the world are spending billions of dollars preparing remote sensor systems to monitor our environment. On the other hand, industry is beginning to develop commercial satellite sensor systems to meet a market-driven demand for primarily high spatial resolution digital image data and has developed GIS systems for a wide variety of commercial and governmental applications – GIS systems that still lack the ability to effectively ingest, adequately manipulate, efficiently query, effectively process, and appropriately manage remotely sensed data.

The authors have, both persistently and consistently, argued that both remote sensing and GIS technologies are intrinsically linked. To be effective, key GIS data layers must be as accurate and up to date as practical. Remote sensing can help update GIS data layers. Remote sensing data often require correlative data to improve the accuracy of given analyses. GIS data layers can be employed to help supply such information. Sometimes we have wondered why the calls that have been made for more funding in this area have fallen on deaf ears. We sincerely believe that the appropriately linked application of remote sensing and GIS can benefit all mankind and can help us achieve the understanding necessary to improve our chances for sustainable economic development.

It is important, then, that we get on with research of the type discussed herein. As stated previously, there are still very serious impediments to the integration of remotely sensed and GIS data. The authors feel that there is a definite need to reassess the overall funding priorities throughout the federal government and industry in both remote sensing and GIS. Research currently underway today is to a very real extent underfunded in the areas of data analysis, technology research and development, and infrastructure and capacity building discussed herein. We all must, by now, realize the critical need for accurate and timely information on the current status and trends of socioeconomic, biophysical, and geochemical processes in our environment. Why aren't we acting on this realization? We understand that there are many pressing issues facing governments and industry today – issues that vie for the resources needed to move research and development forward. We believe, however, that if we do not fund work in this area we will not have the ability to manage our planetary resources as we must.

Remote sensing and GIS offer the only practical means for acquiring and analyzing the data required to develop a path toward sustainable economic development. Such data are urgently needed at scales from local to global. Remote sensing offers us the most cost-effective means for the collection of many types of basic environmental data at any scale above local – data that are internally consistent and whose errors/accuracy we can have any hope of understanding. Yet, to be effective we need to employ GIS to integrate these remotely sensed data with other data, manipulate and perform analyses, extract information, and convey that information to policy makers, environmental planners, and resource managers in an effective fashion. We are, in our opinion, not doing this to the level needed or necessary at this time. This again argues for a reassessment of funding agency and industry priorities.

When governmental agencies spend tens to hundreds of thousands of dollars on GIS procurement only to find that the maps and imagery they wish to input for analysis either do not exist or are outdated, we have a problem. When agencies charged with the production of spatial data are forced to cut way back on the research and development to speed their production processes because they are having difficulty meeting existing needs, we also have a problem. When we spend large amounts of resources on studying issues that may impact humanity thirty to fifty years from now and expend very few resources to help us assess the nature and current status of either local, regional, or national, let alone global land cover and demographic data, we have a problem. We could go on with this listing, but the basic point that we are attempting to make in the starkest of terms is that more resources need to be directed at the development of the production of environmental baseline data. Advanced integrated remote sensing and GIS capabilities are required if we are to accomplish this development in what the authors consider as real-enough time.

Nevertheless, the picture is not totally bleak: There are signs that governmental funding agencies and the legislators they depend upon for their funding in these areas are coming to a new level of realization of the potential of these technologies. Hopefully, this realization will equate to new resources being directed toward research and development activities in the areas described herein, even in this era of government downsizing. We must not consider funding for research in remote sensing/GIS integration a zero sum game. We cannot, and should not, accept that existing funds in remote sensing research areas, in particular, should just be redistributed on the basis of the arguments in support of the research agenda presented here. On the contrary, while we hope that some redirection might occur, the agenda is presented in the hope that it can be used to influence funding agencies and industry to realize the significant benefits we believe can accrue also from directing new resources toward remote sensing and GIS as areas of high priority. In an era when we are trying to do things "better, faster, and cheaper," remote sensing and GIS should be just the ticket.

9.6. A look at the future

The future we see presupposes the existence of a global information infrastructure wherein a key part of the infrastructure is a global spatial data framework. We foresee an information infrastructure where individuals are connected to a wide variety of services, including georeferenced databases via high-speed networks both at home and in the work place; a framework where spatial data, from remote sensor systems and all other collection systems, are acquired at appropriate intervals in the spatial and spectral resolutions appropriate to the uses of the data type; data types whose validity is verified at real, or real-enough, time to the time they are collected; a framework where all these collected data types are integrated in compatible formats in seamless, digital databases. In these databases would reside historical records that have been accumulated and verified to the fullest extent practical. The databases would be well known to the entire digital georeferenced data user community, from the most advanced professional researchers to the general public. These secure, readily accessible digital databases would be entered by administrators and users alike, are quickly verified, are safe from unauthorized tampering, and yet are quickly and easily made available to anyone at no more than the cost of fulfilling a user's request. The databases would facilitate a user's ability to browse, preview, query, select, and acquire a single or multiple representations of integrated spatial datasets at scales from a site to the whole world and everything in between and use object-oriented approaches, where site-based (where site is essentially anything from a plot to the entire globe) datasets can be easily designated and extracted.

We would like to see a future where databases that are accessible via very high speed networks can supply large volumes of spatial data to users in real-enough time. The networks must be able to handle all other forms of data as well, be open, accessible, and low cost with respect to both costs and the overhead required in their use, and capable of taking the data from its source, be that a general archive, a library, an agency, or other institution or an individual, to a user's truly integrated desktop GIS. The desktop GIS must be easy to use and have the widest practical range of image processing and spatial analysis functionality – functionality that combines the ability to address both basic and applied research, development efforts, operational issues, and, yes, even entertainment elements; functionality that assists the less-sophisticated user and enables the more advanced user; functionality that includes a wide range of visualization tools from multidimensional to animation and more; functionality that facilitates the heritage tracking and the validation of output data products; and functionality where data products can be output in a variety of formats. The output can then be readily incorporated in an appropriate fashion back into any database: any database, that is, where the data produced by a user will be of value and can be used to augment and/or update the data in that database. This output will be used by environmental planners, resource managers, and public policy decision makers, used by the private sector, governmental and nongovernmental organizations, the press, and the general public; used by all segments of society to inform and to entertain; used to educate and to protect, used to develop and to conserve, and even used, in particular, to measure, map, monitor, model, and manage our global resource base for the benefit of all mankind.

We have come a long way in the past thirty years, but we still have a long way to go... a lot more to do. This book is dedicated to the memory of my co-author, Jeffrey Star. Jeff, I miss you and I wish you were still here to help.

On the Passing of a Star

He was unique
One of a kind
A friend of the planet
A friend of mankind

So much accomplished
In so little time
So many lives touched
By this friend of mankind

Beloved husband
Caring father of two
From a closely knit family
All so proud of you

You taught us a lot
We'll talk again one day
The time comes to each of us
To follow in your way

Yet somehow we take comfort in knowing
That Jeffrey where ever you are
You know this world is a better place
Because through it passed a Star

J. Estes
A friend

BIBLIOGRAPHY

Adams, P. and Solomon, M., 1993. An overview of the CAPITL software development environment. *Technical Report TR-1 143*, Computer Science Dept., Univ. Wisconsin, Madison.

Adelson, E.H., 1993. Perceptual organization and the judgment of brightness. *Science*, 262:2042–4.

Allder, W., Ziede, A., McEwen, R., and Beck, F., 1984. USGS digital cartographic data standards: digital line graph attribute coding standards. *Circular 895-G USGS*, Reston VA.

Alonso, G. and El Abbadi, A., 1994. Cooperative modeling in applied geographic research. *International Journal of Intelligent and Cooperative Information Systems*, 3(1):83–102.

American Planning Association (APA), 1981. *Had You Planned to be a Planner?* APA, Washington, D.C.

American Society for Photogrammerty and Remote Sensing (ASPRS), 1989. Interim accuracy standards for large scale line maps. *Photogrammetric Engineering and Remote Sensing*, 55(7):1038–40.

Anderson, J., Hardy, E., Roach, J., and Whitmer, R., 1976. A land use and land cover classification for use with remote sensor data. *Geological Survey Professional Paper 964*, U.S. Government Printing Office, Washington, D.C.

Andrews, T., 1991. Using an object database to build integrated design environments. In *Object-Oriented Databases with Applications to CASE, Networks, and VLSI CAD, Data and Knowledge Base Systems*, R. Gupta and E. Horowitz (Eds.), Chapter 17, pp. 313–23, Prentice Hall, Englewood Cliffs, NJ.

Anselin, L., 1992. SPACESTAT, National Center for Geographic Information and Analysis, Santa Barbara, CA.

Aronoff, S., 1982a. Classification accuracy: A user approach. *Photogrammetric Engineering and Remote Sensing*, 48:1299–307.

    1982b. Classification accuracy: A user's view. *Photogrammetric Engineering and Remote Sensing*, 48:1309–12.

Asrar, G. (ed.), 1989. *Theory and Applications of Optical Remote Sensing*. Wiley, New York.

Asrar, G. and D.J. Dokken (ed.), 1993. *EOS Reference Handbook*. NASA Headquarters, Washington, D.C., p. 145.

Atkinson, P., 1991. Optimal ground sampling for remote sensing investigations, estimating the regional mean. *International Journal of Remote Sensing*, 12:559–67.

Augenstein, E.W., Stow, D.A., and Hope, A.S., 1991. Evaluation of SPOT HRV-XS Data for Kelp Resource Inventories. *Photogrammetric Engineering and Remote Sensing*, 57(5):501–9.

Baker, J.R., Marks, G.W., and Mikhail, E.M., 1975. Analysis of digital multispectral scanner (MSS) data. *Bildmessung und Luftbildwesen*, 43:22–27.

Bancilhon, E., Delobel, C., and Kanellakis, P.C., 1992. *Building an Object-Oriented Database System: The Story of O_2*. Morgan Kaufmann, San Mateo, CA.

Beard, M.K., Buttenfield, B.P., and Clapham, S., 1991. Visualizing the quality of spatial information: Scientific report of the Specialist Meeting. *Technical Report 91-26*. National Center for Geographic Information and Analysis, Santa Barbara, CA.

Berners-Lee, J., Cailliau, R., and Pollerman, B., 1992. World Wide Web: The information universe. *Electronic Networking: Research, Applications and Policy*, 1(2):42–58.

Beyer, E.P., 1983. An overview of the Thematic Mapper geometric correction system, Landsat-4 Science Characterization: Early Results, NASA Conference Publication 2355, Vol. II, pp. 87–145.

Billingsley, F.C., Anuta, P.E., Carr, J.L., McGillem, C.D., Smith, D.M., and Strand, T.C., 1983. Data processing and reprocessing. In: R. Colwell (ed.), *Manual for Remote Sensing*, 2nd Ed., American Society for Photogrammetry, Falls Church, VA, Vol. I, pp. 719–92.

Blakemore, M., 1984. Generalization and error in spatial databases. *Cartographica*, 21:131–9.

Boone, C.F., 1989. *And Hugo Was His Name*. Boone Publications, Sun City, AZ, p. 66.

Budd, L., 1990. *GIS for Vermont Communities; Applications and Concepts*. Vermont Office of Geographic Information Services, State of Vermont, Montpelier, VT, p. 128.

Burroughs, D., 1988. The REFIRES (REgional FIre REgime Simulation) model: A C program for regional fire regime simulation. Master's thesis, Dept. Geography, Univ. Calif., Santa Barbara, CA.

Buttenfield, B.P. and McMaster, R.B. (eds.), 1991. *Map Generalization: Making Rules for Knowledge Representation*. Longman House, London.

Cablk, M.R., Kjerfve, B., Michener, W.K., and Jensen, J.R., 1994. Impacts of Hurricane Hugo on a coastal forest: Assessment using Landsat TM data. *Geocarto International: Multidisciplinary Journal of Remote Sensing*, 2:15–24.

Campbell, J., 1981. Spatial correlation effects upon accuracy of supervised classification of land cover. *Photogrammetric Engineering and Remote Sensing*, 47:355–64.

Card, D., 1982. Using known map category marginal frequencies to improve estimates of thematic map accuracy. *Photogrammetric Engineering and Remote Sensing*, 48:431–9.

Castleman, K.R., 1979. *Digital Image Processing*. Prentice-Hall, 429 pp.

Castle, G. (ed.), 1993. *Profiting from a Geographic Information System*. GIS World, Inc., Fort Collins, CO.

Cate, V., 1992. Alex: a global filesystem, *Proceedings of the USENIX File Systems Workshop*, Ann Arbor, MI, USA, May 1992.

Chavez, P., Jr. and Bowell, J., 1988. Comparison of the spectral information content of Landsat Thematic Mapper and SPOT for three different sites in the Phoenix Arizona Region. *Photogrammetric Engineering and Remote Sensing*, 54(12):1699–708.

Chavez, P.S., Jr. and Kwarteng, A.Y., 1989. Extracting spectral contrast in Landsat Thematic Mapper image data using selective principal component analysis. *Photogrammetric Engineering and Remote Sensing*, 55(3):339–48.

Chavez, P.S., Jr., Sides, S.C., and Anderson, J.A., 1991. Comparison of three methods to merge multiresolution and multispectral data: Landsat TM and SPOT Panchromatic, *Photogrammetric Engineering and Remote Sensing*, 57(3):295–303.

Cheng, T.D., Angelici, G.L., Slye, R.E., and Ma, M., 1992. Interactive boundary delineation of agricultural lands using graphics workstations. *Photogrammetric Engineering and Remote Sensing*, 58(10):1439–43.

Chrisman, N.R., 1989. Modeling error in overlaid categorical maps. In *Accuracy of Spatial Databases*, M. Goodchild and S. Gopal (eds.), Taylor & Francis, New York, pp. 21–34.

1991. The error component in spatial data. In *Geographical Information Systems: Principles and Applications*, D.J. Maguire, M.F. Goodchild, and D.W. Rhind (eds.), Longman, New York, pp. 165–74.

Chu, W. (ed.), 1993. *Proceedings of the NSF Scientific Database Projects*, AAAS Workshop on Advances in Data Management for the Scientist and Engineer, Boston, MA, Feb. 1993.

Civco, D.L., 1989. Topographic normalization of Landsat Thematic Mapper digital imagery. *Photogrammetric Engineering and Remote Sensing*, 55(9):1303–9.

Colwell, R. (ed.), 1983. *The Manual of Remote Sensing*, 2nd Ed., American Society of Photogrammetry and Remote Sensing, Fall Church, VA.

Committee on Earth and Environmental Sciences (CEES), 1992. *The U.S. Global Change Data and Information Management Program Plan*. National Science Foundation, Forms and Publication Division, Washington, D.C., p. 94.

Congalton, R.G., 1988. A comparison of sampling schemes used in generating error matrices for assessing the accuracy of maps generated from remotely sensed data. *Photogrammetric Engineering and Remote Sensing*, 54(6):593–600.

Congalton, R., 1988a. Using spatial autocorrelation analysis to explore the errors in maps generated from remotely sensed data. *Photogrammetric Engineering and Remote Sensing*, 54:587–92.

1988b. A comparison of sampling schemes used in generating error matrices for assessing the accuracy of maps generated from remotely sensed data. *Photogrammetric Engineering and Remote Sensing*, 54:593–600.

Congalton, R., Oderwald, R., and Mead, R., 1983. Assessing Landsat classification accuracy using discrete multivariate statistical techniques. *Photogrammetric Engineering and Remote Sensing*, 49:1671–8.

Cooke, D.F., 1980. A review of geoprocessing systems and a look at their future. In *Computers in Local Government Urban and Regional Planning*, K. Krammer and J. King (eds.), Auerback, pp. (2.4.1) 1–16.

Cowen, D.J. and Shirley, W.L., 1991. Integrated planning information systems. In *Geographical Information Systems; Principles and Applications*, Vol. 2, D.J. Maguire, M.F. Goodchild, and D.W. Rhind (eds.), Longman, New York pp. 297–310.

Cowen, D.J., Green, E., Halls, J., Jensen, J.R., and Schmidt, N., 1995. Assessing a commercial real estate market: A case study. *Business Geographics*, 3(6): 38–41.

Craig, R., 1979. Autocorrelation in Landsat data, *Proceedings of the Thirteenth International Symposia on Remote Sensing of Environment*, ERIM, Univ. Michigan, Ann Arbor, pp. 1517–24.

Crist, E. and Cicone, R., 1984. A physically-based transformation of Thematic Mapper data, the TM Tasseled Cap. *IEEE Transactions on Geoscience and Remote Sensing*, GE-22:256–63.

Curran, P. and Hay, A., 1986. The importance of measurement error for certain procedures in remote sensing at optical wavelengths. *Photogrammetric Engineering and Remote Sensing*, 52:229–41.

Dahl, T.W., 1990. *Wetlands Losses in the United States 1780's to 1980's*. U.S. Dept. Interior, U.S. Fish & Wildlife Service, Washington, D.C., p. 21.

Dall, J.A., 1990. The digital orthophoto: The cornerstone of GIS, *Proceedings of GIS/LIS '90*, ASPRS/ACSM, pp. 314–8.

Dangermond, J. and Freedman, C., 1986. Findings regarding a conceptual model for a municipal database and implications for software design. *Geo-Processing*, 3(2):31–49.

Date, C., 1986. *Relational Database: Selected Writings*. Addison-Wesley, Menlo Park, CA.

Davis, F., Quattrochi, D., Ridd, M., Lam, N., Walsh, S., Michaelson, J., Franklin, J., Stow, D., Johannsen, C., and Johnston, C., 1991. Environmental analysis using integrated GIS and remotely sensed data: Some research needs and priorities. *Photogrammetric Engineering and Remote Sensing*, 57:689–97.

de Hoop, S. and van Oosterom, P., 1992. Storage and manipulation of topology in POSTGRES, *Proceedings of EGIS*, pp. 1324–36, Munich, Germany, Mar. 1992.

Dicks, S.E. and Lo, T.H.C., 1990. Evaluation of thematic map accuracy in a land-use and land-cover mapping program. *Photogrammetric Engineering and Remote Sensing*, 56(1):1247–52.

Dobson, J.E., Bright, E.A., Ferguson, R.L., Field, D.W., Wood, L.L., Haddad, K.D., Iredale, H., Jensen, J.R., Klemas, V.V., Orth, R.J., and Thomas, J.P., 1995.

NOAA Coastal Change Analysis Program (C-CAP): Guidance for Regional Implementation, NOAA Technical Report NMFS 123, p. 92.

Dolan, M.G., Martin, S.R., and Warnick, L.J., 1984. Comparative evaluation of simulated SPOT, Landsat TM and NHAP CIR data for urban land cover and impervious surface identification. In *SPOT Simulation Applications Handbook*, American Society of Photogrammetry, Falls Church, VA, pp. 148–56.

Dueker, K.J. and Delacy, P.B., 1990. GIS in the land development planning process; balancing the needs of land use planners and real estate developers. *Journal of the American Planning Association*, Autumn 1990, pp. 483–91.

Duggin, M., 1985. Factors limiting the discrimination and quantification of terrestrial features using remotely sensed radiance. *International Journal of Remote Sensing*, 6:3–27.

Eastman, J.R., 1992. Time series map analysis using standardized principal components, *Proceedings, ASPRS/ACSM Symposium*, ASPRS, Baltimore, MD, pp. 195–204.

Ebner, H., 1979. Zwei neue interpolationsverfahren und beispiele für ihre anwendung. *Bildmessung und Luftbildwesen*, 47:15–27.

Ebner, H., Fritsch, D., and Heipke, C. (eds.), 1991. *Digital Photogrammetric Systems*. Wichmann Verlag, Karlsruhe, p. 344.

Eckhardt, D.W., Verdin, J.P., and Lyford, G.R., 1990. Automated update of an irrigated lands GIS using SPOT HRV imagery. *Photogrammetric Engineering and Remote Sensing*, 56(11):1515–22.

Egels, Y., 1991. *GPS for photogrammetry – experience and projects in France*, Schriftenreihe des Institutes für Photogrammetrie der Universität Stuttgart, No. 15, pp. 79–82.

Ehlers, M., 1984a. The automatic DISCOR system for rectification of space-borne imagery as a basis for map production, *Proceedings of the XVth International Congress of ISPRS*, Rio de Janeiro, Brazil, IV:135–47.

1984b. *Digitale bildverarbeitung*, Schriftenreihe des Institutes für Photogrammetrie, Universität Hannover, No. 9, p. 146.

1987. *Integrative Verarbeitung von digitalen Bilddaten der Satellitenphotogrammetrie und -fernerkundung im Rahmen von Geographischen Informationsmodellen*, Wissenschaftliche Arbeiten der Fachrichtung Vermessungswesen der Universität Hannover, No. 149, p. 137.

1991. *Digitization, digital editing and storage of photogrammetric images*, 43rd Photogrammetric Week, Schriftenreihe des Institutes für Photogrammetrie der Universität Stuttgart, No. 15, pp. 187–93.

1992. Data types and data structures for integrated geographic information systems. In *The Integration of Remote Sensing and Geographic Information Systems*, J.L. Star, (ed.), American Society for Photogrammetry and Remote Sensing, Bethesda, MD, pp. 51–73.

Ehlers, M., Edwards, G., and Bedard, Y., 1989. Integration of remote sensing with geographic information systems; A necessary evolution. *Photogrammetric Engineering and Remote Sensing*, 55(11):1619–27.

Ehlers, M., Jadkowski, M.A., Howard, R.R., and Brostuen, D.E., 1990. Applications of SPOT data for regional growth analysis and local planning. *Photogrammetric Engineering and Remote Sensing*, 56(2):175–80.

Ehlers, M. and Fuller, M., 1991. Region-based matching for image integration in remote sensing databases, *Proceedings IGARRS 1991*, Espoo, Finland, IEEE, New York.

Ehlers, M., Greenlee, D., Smith, T., and Star, J., 1991. Integration of remote sensing and GIS: Data and data access. *Photogrammetric Engineering and Remote Sensing*, 57:669–76.

Estes, J., 1992. Remote sensing and GIS integration: Research needs, status and trends, proceedings ISPRS commission II/VII international workshop on 3D in remote sensing and GIS: Systems and applications. International Institute for Aerial Surveys, *ITC Journal 1992*, No. 1, Enschede, The Netherlands, pp. 2–10.

Estes, J.E., 1992. Technology and policy issues impact global monitoring. *GIS World*, 5(10):52–55.

Estes, J. and Star, J., 1990. *Geographic Information Systems: An Introduction*. Prentice Hall, Englewood Cliffs, NJ, p. 303.

Estes, J. and Star J., 1993. Remote sensing and GIS integration: Towards a prioritized research agenda, *Proceedings Twenty-Fifth International Symposium on Remote Sensing and Global Environmental Change: Tools for Sustainable Development*, April, 1993, Graz, Austria, Vol. I, Ann Arbor, MI, Environmental Research Institute of Michigan, pp. I-448–64.

Estes, J., Star J., and Davis, F., 1995. Integration of remote sensing and GIS: A background to NCGIA Initiative 12, NCGIA, initiative 12, closing report, NCGIA Publication, Univ. Calif., Santa Barbara, p. 53.

Faust, N., 1989. Image Enhancement. In *Encyclopedia of Computer Science and Technology*, Marcel Dekker, New York.

1991. Display technologies for remote sensing/GIS, *Proc. ASPRS Annual Meeting*, Baltimore, MD, March 25–28, American Society of Photogrammetry and Remote Sensing, Fall Church, VA.

Faust, N., Anderson, W., and Star, J., 1991. Geographic information systems and remote sensing future computing environment. *Photogrammetric Engineering and Remote Sensing*, 57:655–68.

Fenstermaker, L. (ed.), 1994. *Remote Sensing Thematic Accuracy Assessment: A Compendium*. American Society for Photogrammetry and Remote Sensing, Bethesda, MD.

Ferguson, R.L., Wood, L.L., and Graham, D.B., 1993. Monitoring spatial change in seagrass habitat with aerial photography. *Photogrammetric Engineering and Remote Sensing*, 59(6):1033–8.

Finlay, M., 1993. *Getting Graphic*. M&T Publishing, San Mateo, CA.

Fisher, L., 1992. Geometric correction of multispectral scanner data using global positioning system and digital terrain models, *Proceedings of a Special Session of the ACSM-ASPRS Annual Convention: The Integration of Remote Sensing and Geographic Information Systems*, J. Star (ed.), American Society for Photogrammetry and Remote Sensing, Falls Church, VA.

Fogel, D. and Tinney, L., 1994. Image registration using multiquadric functions. Final Report for Dept. of Energy P.O. 7870M, Remote Sensing Research Unit, Dept. Geography, Univ. Calif., Santa Barbara, CA.

Foley, J. and Van Dam, A., 1984. *Fundamentals of Interactive Computer Graphics.* Addison-Wesley, Reading, Mass.

Foresman, T., 1984. Mapping, monitoring, and modeling of a hazardous waste site. *The Science of the Total Environment,* 56:255–63.

Foresman, T.W., Kelley, R., and Shalit, H., 1990. GIS insurance: The accuracy question. In *Urban and Regional Information Systems Association Proceedings,* F. Westerlund (ed.), Vol. IV, Edmonton, Alberta, pp. 129–42.

Forshaw, M.R.B., Haskell, A., Miller, P.F., Stanley, D.J., Townshend, J.R.G., 1983. Spatial resolution of remotely sensed imagery; A review paper. *International Journal of Remote Sensing,* 4(3):497–520.

Franklin, J., 1988. Canopy reflectance modeling in a tropical savanna. Ph.D. dissertation, Univ. Calif., Santa Barbara, CA.

Frolov, Y. and Mahling, D., 1969. The accuracy of area measurements by point counting techniques. *Cartographic Journal,* 6:21–35.

Fuller, M.E., and Ehlers, M., 1991. An automated system for image co-registration using interest clump matching. *Technical Papers, 1991 ACSM-ASPRS Annual Convention, Baltimore,* MD, 5:93–102. American Society for Photogrammetry and Remote Sensing, Bethesda, MD.

Fung, T. and LeDrew, E., 1987. Application of principal components analysis for change detection. *Photogrammetric Engineering and Remote Sensing* 53(12):1649–58.

   1988. The determination of optimal threshold levels for change detection using various accuracy indices. *Photogrammetric Engineering and Remote Sensing,* 54(10):1449–54.

Gage, J., 1990. Personal communication. Presentation at NCGIA I-12 Specialists Meeting, EROS Data Center, Sioux Falls, SD.

Gallagher, M.L., 1992. GIS on the job. *Planning,* 58(12):20–23.

Gastellu-Etchegorry, J.P., 1990. An assessment of SPOT XS and Landsat MSS data for digital classification of near urban land cover. *International Journal of Remote Sensing,* 11(2):225–35.

Gernazian, A. and Sperry, S., 1989. The integration of remote sensing and GIS, Advanced Imaging, March 1989, No. 3. pp. 30 33, 59.

Getis, A. and Franklin, J., 1987. Second-order neighborhood analysis of mapped point patterns. *Ecology,* 68:473–7.

Goel, N. and Grier, T., 1986a. Estimation of canopy parameters for inhomogenous vegetation canopies from reflectance data, I. two-dimensional row canopy. *International Journal of Remote Sensing,* 7:665–81.

   1986b. Estimation of canopy parameters for inhomogenous vegetation canopies from reflectance data, II. estimation of leaf area index and percentage ground cover for row canopies. *International Journal of Remote Sensing,* 7:1263–86.

   1988. Estimation of canopy parameters for inhomogenous vegetation canopies from reflectance data, III. TRIM: A model for radiative transfer in heterogeneous three-dimensional canopies. *Remote Sensing of Environment,* 25:255–93.

Goodchild, M., 1980. A fractal approach to the accuracy of geographical measures. *Mathematical Geology*, 12:85–98.

1992. Geographical data modeling. *Computers and Geosciences*, 18:401–8.

Goodchild, M. and Gopal, S. (eds.), 1989. *Accuracy of Spatial Databases*. Taylor & Francis, New York.

Goodchild, M. and Yang, S., 1992. A hierarchical data structure for global geographic information systems. *CVGIP Graphical Models and Image Processing*, 54:31–44.

Goodchild, M., Sun, G., and Yang, S., 1992. Development and test of an error model for categorical data. *International Journal of Geographical Information Systems*, 6:87–104.

Göpfert, W., 1982. Methodology for thematic image processing using thematic and topographic data bases and base-integrated multi-sensor imagery, *Proceedings, ISPRS Commission VII Symposium*, Toulouse, France, Vol. I, pp. 13–19.

Hall, F.G., Strebel, D.E., Nickeson, J.E., and Goetz, S.J., 1991. Radiometric rectification: Toward a common radiometric response among multidate, multisensor images. *Remote Sensing of Environment*, 35:11–27.

Hardy, R.L., 1971. Multiquadric equations of topography and other irregular surfaces. *Journal of Geophysical Research*, 76(8):1905–15.

1990. Theory and applications of the multiquadric-Biharmonic method. *Computers Math. Applic.*, 19(8/9):163–208.

Harel, D., Lachover, H., Naamad, A., Pnuell, A., Politi, M., Sherman, R., Shtull-Trauring, A., and Trakhtenbrot, M., 1990. Statemate: A working environment for the development of complex reactive systems. *IEEE Trans. Sofw. Eng.*, 16(4):403–14.

Harvey, D., 1969. *Explanation in Geography*. Edward Arnold, London, p. 521.

Haydin, R., Dalke, W., and Henkel J., 1982. Application of the IHS transform to the processing of multisensor data and image enhancement. *Proc. International Symposium on Remote Sensing of Arid and Semi-Arid Lands*, Cairo, Egypt, Jan. 1982, Environmental Research Institute of Michigan, Ann Arbor, MI.

Hobbs, B. and Voelker, A., 1978. Analytical multiobjective decision-making techniques and power plant siting: A survey and critique. *Oak Ridge National Laborotories Technical Report ORNL-5288*, NTIS, Springfield, VA.

Holz, R., Huff, D.L., and Mayfield, R.C., 1969. Urban spatial structure based on remote sensing. *Proc. of Sixth Int. Symp. on Remote Sensing of Environ.*, Univ. Michigan, Ann Arbor, pp. 819–30.

Hord, R. and Brooner, W., 1976. Land-use map accuracy criteria. *Photogrammetric Engineering and Remote Sensing*, 42:671–7.

Huete, A., 1984. Separation of soil-plant spectral mixtures by factor analysis. *Remote Sensing of Environment*, 19:237–51.

Hull, R. and King, R., 1987. Semantic data modeling: survey, applications, and research issues. *ACM Comput. Surv.*, 19(3):201–60.

International Geosphere Biosphere Programme (IGBP), 1992. *Improved Global Data for Land Applications*, J. Townshend (ed.), IGBP Global Change Report #20, International Council of Scientific Unions, Stockholm.

Jacob, T., 1991. *System Integration of Inertial Navigation, Satellite Navigation and Laser for Airborne Positioning,* Schriftenreihe des Institutes für Photogrammetrie der Universität Stuttgart, No. 15, pp. 61–72.

Jackson, M.J. and Mason, D.C., 1986. The development of integrated geo-information systems. *International Journal of Remote Sensing,* 7(6):723–40.

Jankowski, P. and ZumBrunnen, C., 1990. A model management approach to modeling and simulation natural systems, *Proceedings of the 4th International Symposium on Spatial Data Handling,* Zurich, Switzerland, International Geographical Union.

Jarvis, P. and McNaughton, K., 1986. Stomatal control of transpiration: scaling up from leaf to region. *Advances in Ecological Research,* 15:1–46.

Jensen, J., 1978. Digital land cover mapping using layered classification logic and physical composition attributes. *The American Cartographer,* 5:121–32.

Jensen, J.R., Ramsey, E.W., Mackey, H.E., Christensen, E.J., and Sharitz, R.R., 1987. Inland wetland change detection using aircraft MSS data. *Photogrammetric Engineering and Remote Sensing,* 53(5):521–9.

Jensen, J.R., Rutchey, K., Koch, M.S., and Narumalani, S., 1995a. Inland wetland change detection in the everglades water conservation area 2A using a time series of normalized remotely sensed data. *Photogrammetric Engineering and Remote Sensing,* 61(2):199–209.

Jensen, J.R., 1995b. Issues involving the creation of digital elevation models and terrain corrected orthoimagery using soft-copy photogrammetry. *Geocarto,* 10(1):5–21.

1996. *Introductory Digital Image Processing: A Remote Sensing Perspective,* 2nd Ed. Prentice-Hall, Englewood Cliffs, NJ, p. 330.

Jensen, J.R. and Narumalani, S., 1992. Improved remote sensing and GIS reliability diagrams, image genealogy diagrams, and thematic map legends to enhance communication. *International Archives of Photogrammetry and Remote Sensing,* 6(B6):125–32.

Jensen, J.R. and Toll, D.L., 1982. Detecting residential land-use development at the urban fringe. *Photogrammetric Engineering and Remote Sensing,* 48(4):629–43.

Jensen, J.R., Bryan, M.L., Friedman, S.Z., Henderson, F.M., Holz, R.K., Lindgren, D., Toll, D.L., Welch, R.A., and Wray, J.R., 1983. Urban/suburban land use analysis. In *Manual of Remote Sensing,* R.N. Colwell (ed.), American Society of Photogrammetry, Falls Church, VA, pp. 1571–666.

Jensen, J.R., Cowen, D.J., Halls, J., Narumalani, S., Schmidt, N.J., Davis, B., and Burgess, B., 1994a. Improved urban infrastructure mapping and forecasting for BellSouth using remote sensing and GIS technology. *Photogrammetric Engineering and Remote Sensing,* 60(3):339–46.

Jensen, J.R., Cowen, D.J., Narumalani, S., Althausen, J.D., and Weatherbee, O., 1993a. An evaluation of coastwatch change detection protocol in South Carolina. *Photogrammetric Engineering and Remote Sensing,* 59(6):1039–46.

Jensen, J.R., Narumalani, S., Weatherbee, O., and Mackey, H.E., 1993b. Measurement of seasonal and yearly cattail and waterlily changes using multidate

SPOT panchromatic data. *Photogrammetric Engineering and Remote Sensing*, 59(4):519–25.

Jensen, J.R., Ramsey, E.W., Mackey, H.E., Christensen, E., and Sharitz, R., 1987. Inland wetland change detection using aircraft MSS data. *Photogrammetric Engineering and Remote Sensing*, 53(5):521–29.

Jones, C., 1990. An introduction to graph-based modeling systems, Part 1: Overview. *ORSA Journal on Computing*, 2(2):136–51.

1991. An introduction to graph-based modeling systems, Part H: Graph-grammars and the implementation. *ORSA Journal on Computing*, 3(3):180–206.

1993. An integrated modeling environment based on attributed graphs and graph-grammars. *Decision Support Systems*, 10:255–75.

Journel, A., 1989. *Fundamentals of Geostatistics in Five Lessons.* American Geophysical Union, Washington, D.C.

Jupp, D., Strahler, A., and Woodcock, C., 1988. Autocorrelation and regularization in digital images I. Basic theory. *IEEE Trans. on Geoscience and Remote Sensing*, 26:463–73.

1989. Autocorrelation and regularization in digital images II. Simple image models. *IEEE Trans. on Geoscience and Remote Sensing*, 27:247–58.

Kaufman, Y. and Fraser, R., 1984. Atmospheric effect on classification of finite fields. *Remote Sensing of Environment*, 15:95–118.

Kauth, R. and Thomas, G., 1976. The Tassled Cap – A graphic description of the spectral-temporal development of agricultural crops as seen by Landsat. *Proc. Symposium on Machine Processing of Remotely Sensed Data*, LARS, Purdue University, West Lafayette, IN.

Keefer, B.J., Smith, J.L., and Gregoire, T.G., 1988. Simulating manual digitizing error with statistical models. *Proceedings, GIS/LIS 88*, Bethesda, MD: ASPRS/ACSM.

Kennedy, M. and Guinn, C., 1975. *Automated Spatial Data Information Systems: Avoiding Failure.* Urban Studies Center, Louisville, KY.

Kerekes, J. and Landgrebe, D., 1991. An analytical model of Earth-observational remote sensing systems. *IEEE Transactions on Geoscience and Remote Sensing*, 21:125–33.

Khorram, S., Bening, G., Chrisman, N., Colby, D., Congalton, R., Dobson, J., Ferguson, R., Goodchild, M., Jensen, J., and Mace, T., 1996. Remote sensing change detection issues and error evaluation. *Photogrammetric Engineering and Remote Sensing*, Peer-Reviewed Monograph, in review.

Kidner, D., Jones, C.B. Knight, D.G., and Smith, D.H., 1990. Digital terrain models for radio path profiles. *Proceedings of the 4th International Symposium on Spatial Data Handling*, Zurich, Switzerland, International Geographical Union.

Kimes, D., Smith, J., and Ranson, K., 1980. Vegetation reflectance measurements as a function of solar zenith angle. *Photogrammetric Engineering and Remote Sensing*, 46(12):1563–73.

Kneizys, F.X., Shettle, E.P., Abreu, L.W., Chetwynd, J.H., Anderson, G.P., Gallery, W.O., Selby, J.E.A., and Clough, S.A., 1988. *User's Guide to LOWTRAN7.* Air Force Geophysics Laboratory Report AFGL-TR-88-0177, Bedford, MA.

Konecny, G., 1976. *Mathematical models and procedures for the geometric restitution of remote sensing imagery*, Schriftenreihe des Institutes für Photogrammetrie, Universität Hannover, No. 1, p. 33.

   1979. Methods and possibilities for digital differential rectification. *Photogrammetric Engineering and Remote Sensing*, 45(6):727–34.

Konecny, G. and Lehmann, G., 1984. *Photogrammetrie*. De Gruyter, Berlin, p. 384.

Konecny, G., Lohmann, P., Engel, H., and Kruck, E., 1987. Evaluation of SPOT imagery on analytical photogrammetric instruments. *Photogrammetric Engineering and Remote Sensing*, 53:1223–30.

Kruse, F., Calvin, W., and Siznec, O., 1988. Automated extraction of absorption features from Airborne Visible/Infrared Imaging Spectrometer (AVIRIS) and Geophysical and Environmental Research Imaging Spectrometer (GERIS) data, *Proceedings of the Airborne Visible/Infrared Imaging Spectrometer (AVIRIS) Performance Workshop*, June 6–8, 1988, G. Vane (ed.), NASA Jet Propulsion Laboratory, Pasadena, CA.

Kumler, M., 1992. An intensive comparison of TINs and DEMs. Ph.D. dissertation, Dept. Geography, Univ. Calif., Santa Barbara, CA.

Labovitz, M., 1984. The influence of autocorrelation in signature extraction – An example from a geobotanical investigation of the Cotter Basin, Montana. *International Journal of Remote Sensing*, 5:315–32.

Lanter, D., 1989. Techniques and method of spatial database lineage tracking. Ph.D. Dissertation, Univ. South Carolina.

Lanter, D. and Veregin, H., 1990. A lineage meta-database program for propagating error in geographic information systems. *Proceedings of GIS/LIS, 1990*, Anaheim, CA, ASPRS.

Lauer, D., Estes, J., Jensen, J., and Greenlee, D., 1991. Institutional issues affecting the integration and use of remote sensing data and geographic information systems. *Photogrammetric Engineering and Remote Sensing*, 57:647–54.

Legendre, P. and Fortin, M., 1989. Spatial pattern and ecological analysis. *Vegetatio* 80:107–38.

Li, X. and Strahler, A., 1985. Geometric-optical modeling of a conifer forest canopy, *IEEE Trans. on Geoscience and Remote Sensing*, GE-23:705–21.

Lillisand, T. and Kiefer, R., 1987. *Remote Sensing and Image Interpretation*. 2nd ed., Wiley, New York.

Lindstrom, P., Koller, D., Hodges, L., Ribarsky, W., Faust, N., and Turner, G., 1995. Level-of-detail management for real-time rendering of phototextured terrain, Georgia Tech. *GVU Report GT-GVU-95-10*, Atlanta, GA.

Liu, L., 1991. Object database support for CASE. In *Object-Oriented Data Bases with Applications to CASE, Networks, and VLSI CAD, Data and Knowledge Base Systems*, R. Gupta and E. Horowitz (Eds.), Chapter 14, pp. 261–82, Prentice Hall, Englewood Cliffs, NJ.

Long, D., Mantey, P.E., Pang, A.T., Langdon, G.G., Jr, Levinson, R.A., Kolsky, H.G., Oritton, B.R., Wash, C.H., and Rosenfeld, L.K., 1992. REINAS: Real time environmental information network and analysis system: concept statement. *Technical Report UCSC-CRL-93-05.* Baskin Center for Computer & Information Sciences, U.C. Santa Cruz.

Lovejoy, S. and Schertzer, D., 1988. Extreme variability, scaling and fractals in remote sensing: analysis and simulation. In *Digital Image Processing in Remote Sensing*, J.-P. Muller (ed.), Taylor & Francis, London, pp. 177–213.

Ludwig, J., 1979. A test of different quadrat variance methods for the analysis of spatial pattern. In *Spatial and Temporal Analysis in Ecology*, R. Cormack and J. Ord (eds.), International Cooperative Publishing House, Fairland, MD, pp. 289–304.

Lu, Yun-Chi, 1992. *Output Data Products and Input Requirements, Version 2.0, Earth Observing System. Vol. I: Instrument Data Product Characteristics,* Science Processing Support Office, NASA Goddard Space Flight Center, Greenbelt, MD.

Maas, S., 1988. Use of remotely-sensed information in agricultural growth models. *Ecological Modelling*, 41:247–68.

Maffini, G., Arno, M., and Bitterlich, W., 1989. Observations and comments on the generation and treatment of error in digital GIS data. In *Accuracy of Spatial Databases*, Taylor & Francis, New York, pp. 55–68.

Maguire, D.J., Goodchild, M.F., and Rhind, D.W. (eds.). 1991. *Geographical Information Systems; Principles and Applications*, Vols. 1 & 2, Longman, New York, p. 1046.

Mahling, D., 1989. *Measurements from Maps: Principles and Methods of Cartometry.* Pergammon Press, New York.

Maillard, P. and Cavayas, F., 1989. Automatic map-guided extraction of roads from SPOT imagery for cartographic database updating. *International Journal of Remote Sensing*, 10(11):1775–87.

Maling, D.H., 1991. Coordinate systems and map projections for GIS. In *Geographical Information Systems: Principles and Applications*, D.J., Maguire, M.F. Goodchild, and D.W. Rhind (eds.), Longman, London, Vol. I, pp. 135–46.

Marble, D. and Peuquet, D., 1983. Geographic information systems and remote sensing. In *Manual of Remote Sensing*, R. Colwell, D. Simonett, J. Estes (eds.), American Society of Photogrammetry, Bethesda, MD, pp. 923–58.

Maynard, P. and Strahler, A., 1981. The logit classifier: A general maximum likelihood discriminant for remote sensing applications. *Proceedings of the 15th International Symposium on Remote Sensing of Environment*, ERIM, Ann Arbor, MI.

McFarland, M., Miller, R., and Neale, C., 1990. Land surface temperature derived from the SSM/I passive microwave brightness temperatures. *IEEE Trans. Geoscience and Remote Sensing*, 28(5):829–38.

McGwire, K. and Estes, J., 1987. Interpolation and uncertainty in GIS modeling. *Proceedings of the International Geographic Information Systems Symposium*, Crystal City, VA, Nov. 15–18, 1987, NASA, Washington, D.C.

McGwire, K., Friedl, M., and Estes, J., 1993. Spatial structure, sampling, and scale in a California savanna woodland. *International Journal of Remote Sensing*, 14:2137–64.

McHarg, I.L., 1969. *Design with Nature.* Natural History Press, Garden City, NY, p. 197.

McKeown, D., Harvey, W., and McDermott, J., 1984. Rule-based interpretation of aerial imagery. *IEEE Transactions on Pattern Analysis and Machine Intelligence*, PAMI-7:570–85.

Mead, D., 1982. Assessing data quality in geographic information systems. In *Remote Sensing for Resource Management*, C. Johannsen and J. Sanders (eds.), Soil Conservation Society of America, Ankeny, IA.

Meaille, R. and Wald, L., 1990. Using geographical information system and satellite imagery within a numerical simulation of regional urban growth. *International Journal of Geographical Information Systems*, 4(4):445–56.

Medeiros, C. and Pires, E., 1994. Databases for GIS. *SIGMOD Record*, 23(1):107–15.

Middelkoop, H. and Janssen, L.L.F., 1991. Implementation of temporal relationships in knowledge based classification of satellite images. *Photogrammetric Engineering and Remote Sensing*, 57(7):937–45.

Mikhail, E.M., Akey, M.L., and Mitchell, O.R., 1984. Detection and sub-pixel location of photogrammetric targets in digital images. *Photogrammetria*, 39:63–83.

Millette, T.L., 1989. An expert system design for remote sensing application strategy development. *Proceedings of ASPRS/ACSM Annual Meeting*, Baltimore, MD, Vol. 3, pp. 24–30.

1990. The Vermont GIS: A model for using regional planning commissions to deliver GIS in support of growth management. In *Geographic Information Systems; Developments and Applications*, L. Worrall (ed.), Belhaven Press, London, UK, pp. 65–86.

1992. Vermont planners add image processing to GIS tools. *Geo Info Systems*, 2(5):42–5.

Miller, J., Wu, J., Boyer, M., Belander, M., and Hare, E., 1991. Seasonal patterns in leaf reflectance red-edge characteristics. *International Journal of Remote Sensing*, 12(7):1509–23.

Millette, T.L. and Sickley, T., 1992. Integration of Idrisi raster land cover classifications with Vermont 1:5000 orthophoto based Arc/Info coverages. *VGIS Technical Paper #7*, Vermont Center for Geographic Information, UVM, Burlington, VT, p. 9.

Monkhouse, F. and Wilkinson, H., 1973. *Maps and Diagrams: Their Compilation and Construction*. Methuen, London.

Montgomery, D. and Peck, E., 1982. *Introduction to Linear Regression Analysis*. Wiley, New York.

Moore, D. and Keddy, P., 1989. The relationship between species richness and standing crop in wetlands: The importance of scale. *Vegetatio*, 79:99–106.

Morris, D. and Flavin, R., 1990. A digital terrain model for hydrology, *Proceedings of the 4th International Symposium on Spatial Data Handling*, Zurich, Switzerland, International Geographical Union.

Müller, J.C., 1991. Generalization of spatial databases. In *Geographical Information Systems: Principles and Applications*, D.J, Maguire, M.F. Goodchild, and D.W. Rhind (eds.), Longman, London, Vol. I, pp. 457–75.

National Aeronautics and Space Administration (NASA), 1987. *HIRIS Instrument Panel Report*, National Aeronautics and Space Administration, Washington, D.C.

National Research Council (NRC), 1990. *Spatial Data Needs: The Future of the National Mapping Program*, National Academy Press, Washington, D.C., p. 78.

National Aeronautics and Space Administration (NASA), 1992. *EOS Data and Information System*. NASA Headquarters, Washington, D.C., p. 31.

National Aeronautics and Space Administrations (NASA), 1993. *Mission to Planet Earth: Catalog of Education Programs and Activities*. NASA Headquarters, Washington, D.C., p. 40.

Nellis, M.D., Lulla, K., and Jensen, J., 1990. Interfacing geographic information systems and remote sensing for rural land use analysis. *Photogrammetric Engineering and Remote Sensing*, 56(3):329–31.

Neumann, B, 1992. PROSPERO: A tool for organizing internet resources. *Electronic Networking: Research, Applications, and Policy*, 2(1).

Newcomer, J. and Szajgin, J., 1984. Accumulation of thematic error in digital overlay analysis. *The American Cartographer*, 11:58–62.

Novak, K., 1992. Rectification of digital imagery. *Photogrammetric Engineering and Remote Sensing*, 58(3):339–44.

Nystrom, D., Wright, B., Prisloe, M., and Batten, L., 1986. USGS/Connecticut geographic information system project. *Technical Papers, 1986 ACSM-ASPRS Annual Convention*, Vol. 3, Geographic Information Systems, American Society for Photogrammetry and Remote Sensing, Bethesda, MD.

Oarg, P. and Harrison, A., 1990. Quantitative representation of land-surface morphology from digital elevation models, *Proceedings of the 4th International Symposium on Spatial Data Handling*, Zurich, Switzerland, International Geographical Union.

Obraczka, K., Danzig, P., and Li, S., 1993. Internet resource discovery services. *Computer*, 26(9):8–22.

Openshaw, S. and Taylor, P., 1981. The modifiable unit area problem. In *Quantitative Geography: A British View*, N. Wrigley and R. Bennett (eds.), Routledge, London.

Paine, D., 1981. *Aerial Photography and Image Interpretation for Resource Management*. Wiley, New York.

Parent, P. and Church, R., 1988. Evolution of geographic information systems as decision making tools, San Francisco: *GIS '87*, pp. 63–71, American Society for Photogrammetry and Remote Sensing, Bethesda, MD.

Parker, H.D., 1988. The unique qualities of a geographic information system; A commentary. *Photogrammetric Engineering and Remote Sensing*, 54(11): 1547–9.

Paul, H., Schek, H.J., Scholl, M.H., Weikum, G., and Deppisch, U., 1987. Architecture and implementation of the Darmstadt database kernel system. *Proc. ACM SIGMOD Int. Conf. on Management of Data*, pp. 196–206, San Francisco, CA, May 1987.

Peckham, J. and Maryanski, F., 1988. Semantic data models. *ACM Comput. Surv.*, 20(3):153–89.

Perkal, J., 1956. On epsilon length. *Bulletin de l'Academie Polonaise des Sciences*, 4:399–403.

1966. On the length of empirical curves. Discussion Paper #10, Michigan Inter-University Community of Mathematical Geographers, Ann Arbor, MI.

Peucker, T., 1976. A theory of the cartographic line. *International Yearbook of Cartography*, 16:134–43.

Peuquet, D.J., 1977. *Scanning, Processing, and Plotting of Cartographic Documents.* Buffalo, NY, Geographic Information Systems Laboratory, State University of New York, p. 122.

Plunk, D.E., Jr., Morgan, K., and Newland, L., 1990. Mapping impervious cover using Landsat TM data. *Journal of Soil and Water Conservation*, 45(5):589–91.

Price, K.P., Pyke, D.A., and Mendes, L., 1992. Shrub dieback in a semiarid ecosystem: The integration of remote sensing and GIS for detecting vegetation change. *Photogrammetric Engineering and Remote Sensing*, 58(4):455–63.

Prince, S. and Tucker, C., 1986. Satellite remote sensing of rangelands in Botswana, II: NOAA AVHRR and herbaceous vegetation. *International Journal of Remote Sensing*, 7:1555–70.

Ramakrishnan, R., Srivastava, D., and Sudarshan, S., 1992. CORAL – control, relations and logic. In *Proceedings of the 18th International Conference on Very Large Data Bases*, L.-Y. Yuan (Ed.), pp. 238–50, Vancouver, BC, Canada, Aug. 1992, Morgan Kaufmann Publishers.

Rao, H. and Peterson, L., 1993. Accessing files in an internet: The jade file system. *IEEE Transactions on Software Engineering*, 19(6):613–24.

Rhyne, T.M., 1992. Using visualization tools at the U.S. environmental protection agency visualization center, Joint U.S. Geological Survey/Jet Propulsion Laboratory Scientific Visualization Workshop, Norfolk VA, May 18–19, 1992, *USGS Open File Report 92-285.*

Richards, J.A., 1986. *Remote Sensing Digital Image Analysis.* Springer Verlag, Berlin, p. 281.

Rosenfield, G. and Fitzpatrick-Lins, K., 1986. A coefficient of agreement as a measure of thematic classification accuracy. *Photogrammetric Engineering and Remote Sensing*, 52:223.

Roth, K., 1991. Practical problems for developing rules in automated mapping. *Technical Papers, 1991 ACSM-ASPRS Annual Convention*, 2:287–92, American Society for Photogrammetry and Remote Sensing, Bethesda, MD.

Rowe, L. and Stonebraker, M., 1987. The POSTGRES data model. *Proceedings of the 13th Conference on Very Large Data Bases*, Brighton, England, 1987, pp. 83–96.

Running, S., Loveland, T., and Pierce, L., 1994. A vegetation classification logic based on remote sensing for use in global biogeochemical models. *Ambio* 12(1):77–81.

Rutchey, K. and Velcheck, L., 1994. Development of an everglades vegetation map using a SPOT image and the global positioning system. *Photogrammetric Engineering and Remote Sensing*, 60:767–75.

Sachs, L., 1974. *Angewandte Statistik.* Springer Verlag, Berlin, p. 548.

Sader, S.A. and Winne, J.C., 1992. RGB/NDVI color composites for visualizing forest change dynamics. *International Journal of Remote Sensing*, 13: 3055–67.

Sakashita, S. and Tanaka, Y., 1989. Computer-aided drawing conversion; An interactive approach to digitize maps, *Proceedings of GIS/LIS '89,* Orlando, FL, Vol. 2, ASPRS/ACSM, Bethesda, MD, pp. 578–90.

Saran, A., Park, K., Chen, Y., Agniar, A., Smith, T.R., and Su, J., 1993. Developing applications in CORAL. In *International Logic Programming Symposium-Workshop on Programming in Deductive Databases,* Springer-Verlag, New York.

Sastri, A., 1994. Experiences in heterogeneous service and data access using distributed approaches. Master's thesis, Comp. Sci. Dept., Univ. Calif., Santa Barbara.

Scholl, M. and Voisard, A., 1989. Thematic map modeling. In *Design and Implementation of Large Spatial Databases: First Symposium,* Vol. 409 of *Lecture Notes in Computer Science.* Springer-Verlag, New York, pp. 167–92.

Schowengerdt, R., Park, S., and Gray, R., 1984. Topics in two-dimensional sampling and reconstruction of images. *International Journal of Remote Sensing,* 5(2):333–47.

Schuhr, W., 1982. *Geometrische Verarbeitung Multispektraler Daten von Zeilenabtastern,* Wissenschaftliche Arbeiten der Fachrichtung Vermessungswesen der Universität Hannover, No. 115, p. 198.

Schultink, G., 1982. Integrated remote sensing and information management procedures for agricultural production potential assessment and resource policy design in developing countries. *Third Asian Conference on Remote Sensing,* Dec. 4–7, 1982. Dacca, Bangladesh.

Sellars, P., 1985. Canopy reflectance, photosynthesis, and transpiration. *International Journal of Remote Sensing,* 6:1335–72.

Shanks, R. and Wang, S., 1992. Integrating digital orthophotography and GIS: A software-based approach to vector/raster processing. *Proceedings of GIS/LIS'92,* ARPRS/ACSM, pp. 674–82.

Shelton, R. and Estes, J., 1981. Remote sensing and geographic information systems: An unrealized potential. *Geo-processing* 1:395–420.

Showengert, R., 1983. *Techniques for Image Processing and Classification in Remote Sensing.* Academic Press, Orlando, FL.

Shy, I., Taylor, R., and Osterweil, L., 1989. A metaphor and a conceptual architecture for software development environments. In *Software Engineering Environments. Proc. Int. Workshop on Environments,* F. Long (ed.), Chinon, France, Vol. 467 of *Lecture Notes in Computer Science,* Springer-Verlag, New York, pp. 77–97.

Siggraph '89, 1989. *Proceedings of the 16th Annual Conference on Computer Graphics and Interactive Techniques,* July 31–Aug. 4, Boston, MA, 1991. ACM Press, New York.

Simonett, D.S., Reeves, R.G., Estes, J.E., Bertke, S.E., and Sailer, C.T., 1983. The development and principles of remote sensing. In *Manual of Remote Sensing,* R.N. Colwell (ed.), American Society of Photogrammetry, Falls Church, VA, pp. 1–32.

Sinton, D. and Steinitz, C., 1971. *Grid Manual, Version 3,* Laboratory for Computer Graphics and Spatial Analysis, Harvard University, Cambridge, MA.

Skidmore, A. and Turner, B., 1988. Forest mapping accuracies are improved using a supervised non-parametric classifier with SPOT data. *Photogrammetric Engineering and Remote Sensing*, 54:1415–21.

Slama, C. (ed.), 1980. *Manual of Photogrammetry*. American Society of Photogrammetry and Remote Sensing, Fall Church, VA.

Smith, E., Oh, K., and Smith, M.,1989. A PC-based interactive imaging system designed for INSAT data analysis and monsoon studies. *Bull. Amer. Meteorol. Soc.*, 70(9):1105–22.

Smith, T., Su, J., El Abbadi, A., Agrawal, D., Alonso, G., and Saran, A., 1994a. Computational modeling systems: Support for the development of scientific models. Technical Report, Comp. Sci. Dept., Univ. Calif., Santa Barbara.

Smith, T., Su, J., and Saran, A., 1994b. Virtual structures – A technique for supporting scientific database applications. *Proc. 13th Int. Conf. on the E-R Approach*, Springer-Verlag, New York.

Solomon, B. and Haynes, K., 1984. A survey and critique of multiobjective power plant siting decision rules. *Socio-Economic Planning Sciences*, 18:71–9.

Song, X. and Osterweil, L., 1994. Engineering software design processes to guide process execution. *Technical Report 94-23*, Comp. Sci. Dept., Univ. Mass.

Star, J., Estes J., and Davis, F., 1995. Research Initiative 12: Integration of remote sensing and geographic information systems, closing report, National Center for Geographic Information and Analysis, Santa Barbara, CA.

Star, J. and Estes, J.E., 1990. *Geographic Information Systems; An Introduction*. Prentice Hall, Englewood Cliffs, NJ, p. 303.

Steiner, D.R., 1991. The integration of digital orthophotographs with GIS's in a microcomputer environment. *ITC-Journal* 1992-1, pp. 65–72.

Stevens, W., 1992. *Advanced Programming in the UNIX Environment*. Addison-Wesley, Reading, Mass.

Stoms, D., 1992. Mapping and monitoring regional patterns of species richness from geographic information. Ph.D. dissertation, Dept. Geography, Univ. Calif., Santa Barbara, CA.

Stonebraker, M., Rowe, L.A., and Hirohama, M., 1990. The implementation of POSTGRES. *IEEE Trans. on Knowledge and Data Engineering*, 2:125.

Stoney, W.E., 1995. Personal Correspondence, Washington: Mitre Corp.

Story, M. and Congalton, R., 1986. Accuracy assessment, A user's perspective. *Photogrammetric Engineering and Remote Sensing*, 52:397–9.

Stow, D., Westmoreland, S., McKinsey, D., Collins, D., Mertz, F., Nagel, D., and Sperry, S., 1990. Raster-vector integration for updating land use data. *Proceedings of the Twenty-Third International Symposium on Remote Sensing of the Environment*, Bangkok, Thailand, April 1990, Vol. 2, pp. 837–44.

Strahler, A.H., Woodcock, C.E., and Smith, J.A., 1986. On the nature of models in remote sensing. *Remote Sensing of the Environment*, 20(2):121–39.

Streich, T.A., 1986. Geographic data processing: A contemporary overview. Master's thesis, Dept. Geography, Univ. Calif., Santa Barbara.

Swanberg, N. and Peterson, D., 1987. Using the airborne imaging spectrometer to determine nitrogen content in coniferous forest canopies. *Proceedings of the International Geoscience and Remote Sensing Symposium*, IEEE, New York.

Systems Development Corp., 1968. *Urban and Regional Information Systems: Support for Planning in Metropolitan Areas.* U.S. Dept. Housing and Urban Development, Washington, D.C., p. 100.

Taylor, R., Belz, F., Clarke, L., Osterweil, L., Selby, R., Wileden, J., Wolf, A., and Young, M., 1988. Foundations for the Arcadia environment architecture. *Technical Report 88-96*, Comp. Sci. Dept., Univ. Mass.

Thapa, K. and Bossler, J., 1992. Accuracy of spatial data used in geographic information systems. *Photogrammetric Engineering and Remote Sensing*, 58(6):835–41.

Thormodsgard, J.M. and Feuquay, J.W., 1987. Larger scale image mapping with SPOT. *Proceedings of the SPOT-1 Utilization and Assesment Results*, Paris France, SPOT Image Corp., Toulouse, France.

Thorpe, J., 1991. *The future of digital orthophotography in GIS in the USA*, Schriftenreihe des Institutes für Photogrammetrie der Universität Stuttgart, No. 15, pp. 145–53.

Times Atlas of the World, 1981. Times Books, London.

Toll, D., 1984. An evaluation of simulated thematic mapper data and Landsat MSS data for discriminating suburban and regional land use and land cover. *Photogrammetric Engineering and Remote Sensing*, 50:1713.

Ton, J. and Jain, A.K., 1989. Registering Landsat images by point matching. *IEEE Transactions on Geosciences and Remote Sensing*, 27(5):642–51.

Townshend, J.R.G. and Justice, C.O., 1988. Selecting the spatial resolution of satellite sensors required for global monitoring of land transformations. *International Journal of Remote Sensing*, 9:187–236.

Townshend, J. and Justice, C., 1990. The spatial variation of vegetation changes at very coarse scales. *International Journal of Remote Sensing*, 11:149–57.

Treitz, P.M., Howarth, P.J., and Gong, P., 1992. Applications of satellite and GIS technologies for land cover and land use mapping at the rural-urban fringe: A Case Study. *Photogrammetric Engineering and Remote Sensing*, 58(4): 439–48.

Tucker, C., 1979. Red and photographic infrared linear combinations for monitoring vegetation. *Remote Sensing of Environment*, 8:127–50.

Tucker, C., Vanpraet, C., Boerwinkel, E., and Gaston, A., 1983. Satellite remote sensing of total dry matter production in the Sengalese Sahel. *Remote Sensing of Environment*, 13:461–74.

Turner, M., Dale, V., and Gardner, R., 1989a. Predicting across scales: theory, development, and testing. *Landscape Ecology*, 3: 245–52.

Turner, M., O'Neill, R., Gardner, R., and Milne, B., 1989b. Effects of changing spatial scale on the analysis of landscape. *Landscape Ecology*, 3:153–62.

United States Geological Survey (USGS), 1992. Joint USGS/Jet Propulsion Laboratory Scientific Visualization Workshop. *USGS Open File Report 92-285*, Washington, D.C.

van Oosterom, P. and Vijlbrief, T., 1991. Building a GIS on top of the open DBMS POSTGRES. *Proceedings of EGIS*, Brussels, Belgium, April 1991, EGIS Foundation, Netherlands, pp. 775–87.

Ventura, A.D., Rampini, A., and Schettini, R., 1990. Image registration by recognition of corresponding structures. *IEEE Transactions on Geosciences and Remote Sensing*, 27(5):642–51.

Veregin, H., 1989. A taxonomy of error in spatial databases. *Technical Paper 89-12*, National Center for Geographic Information and Analysis, Santa Barbara, CA.

Vitec, J., Walsh, S., and Gregory, M., 1984. Accuracy in geographic information systems: An assessment of inherent and operational errors, *Proceedings, PECORA IX Symposium*, pp. 296–302.

Walsh, S., Lightfoot, D., and Butler, D., 1987. Recognition and assessment of error in geographic information systems. *Photogrammetric Engineering and Remote Sensing*, 53:1423–30.

Wang, F., 1990a. Improving remote sensing image analysis through fuzzy information representation. *Photogrammetric Engineering and Remote Sensing*, 56(8):1163–69.

1990b. Fuzzy supervised classification of remote sensing images. *IEEE Transactions on Geoscience and Remote Sensing*, 28(2):194–201.

Wang, J., 1992. Road network detection from SPOT imagery for updating geographic information systems in the rural-urban fringe. *International Journal of Geographical Information Systems*, 6(2):141–57.

Waters, R.S., 1989. Data capture for the nineties; VTRAK, *Proceedings of AUTOCARTO 9*, ACSM/ASPRS, Bethesda, MD, pp. 377–83.

Waterfeld, W. and Schek, H., 1992. The DASDBS Geo-Kernel – An extensible database system for GIS. In *Three Dimensional Modeling with Geoscientific Information Systems*, A.K. Turner (ed.), Chapter 8, pp. 69–84. Kluwer Academic, Dordrecht, Netherlands.

Weibel, R. and Heller, M., 1990. A framework for digital terrain modeling. *Proceedings of the 4th International Symposium on Spatial Data Handling*, Zurich, Switzerland, International Geographical Union.

Welch, R., Jordan, T., and Ehlers, M., 1985. Comparative evaluations of the geodetic accuracy and cartographic potential of Landsat-4 and Landsat-5 Thematic Mapper image data. *Photogrammetric Engineering and Remote Sensing*, 51:1249–62.

Welch, R., 1987. Integration of photogrammetric, remote sensing and database technologies for mapping applications. *Photogrammetric Record*, 12(70):409–28.

Welch, R. and Ehlers, M., 1987. Merging multiresolution SPOT HRV and Landsat TM data. *Photogrammetric Engineering and Remote Sensing*, 53(3):301–3.

Welch, R., Jordan, T.R., and Ehlers, M., 1985. Comparative evaluations of the geodetic accuracy and cartographic potential of Landsat-4/-5 TM image data. *Photogrammetric Engineering and Remote Sensing*, 51(9):1249–62.

Westmoreland, S. and Stow, D.A., 1992. Category identification of changed land-use polygons in an integrated image processing/geographic information system. *Photogrammetric Engineering and Remote Sensing*, 58(11):1593–99.

Whitmore, G.D., 1952. The development of photogrammetry. In *Manual of Photogrammetry*, American Society of Photogrammetry, Washington, D.C., pp. 1–16.

Williamson, H., 1989. The discrimination of irrigated orchard and vine crops using remotely sensed data. *Photogrammetric Engineering and Remote Sensing*, 55:77–82.

Wolf, A., 1989. The DASDBS GEO-Kernel, concepts, experiences, and the second step. In *Design and Implementation of Large Spatial Databases: First Symposium*, Vol. 409 of *Lecture Notes in Computer Science*. Springer-Verlag, New York. pp. 67–88.

  1990. How to fit Geo-objects into databases – An extensibility approach, *Proceedings of the First European Conference on GIS*, Amsterdam, April 1990.

Woodcock, C. and Strahler, A., 1987. The factor of scale in remote sensing. *Remote Sensing of Environment*, 21:311–32.

Zdonik, S. and Maier, D. 1990. *Readings in Object-Oriented Database Systems*. Morgan Kaufmann, San Mateo, CA.

Zhou, Q., 1989. A method for integrating remote sensing and geographic information systems. *Photogrammetic Engineering and Remote Sensing*, 55(5):591–6.

Index

Printed in the United States
By Bookmasters